Active Subspaces

SIAM Spotlights

SIAM Spotlights is a new book series that comprises brief and enlightening books on timely topics in applied and computational mathematics and scientific computing. The books, spanning 125 pages or less, will be produced on an accelerated schedule and will be attractively priced.

Josef Málek and Zdeněk Strakoš, *Preconditioning and the Conjugate Gradient Method in the Context of Solving PDEs*

Paul G. Constantine, *Active Subspaces: Emerging Ideas for Dimension Reduction in Parameter Studies*

Active Subspaces

*Emerging Ideas for Dimension
Reduction in Parameter Studies*

Paul G. Constantine

Colorado School of Mines
Golden, Colorado

Society for Industrial and Applied Mathematics
Philadelphia

Publisher	David Marshall
Acquisitions Editor	Elizabeth Greenspan
Developmental Editor	Gina Rinelli
Managing Editor	Kelly Thomas
Production Editor	Lisa Briggeman
Copy Editor	Nicola Cutts
Production Manager	Donna Witzleben
Production Coordinator	Cally Shrader
Compositor	Scott Collins
Graphic Designer	Lois Sellers

Library of Congress Cataloging-in-Publication Data

Constantine, Paul G.
 Active subspaces : emerging ideas for dimension reduction in parameter studies / Paul G. Constantine, Colorado School of Mines, Golden, Colorado.
     pages cm. -- (SIAM spotlights ; 2)
 Includes bibliographical references and index.
 ISBN 978-1-611973-85-3
 1. Function spaces. 2. Functional analysis. 3. Shift operators (Operator theory) 4. Hilbert space. 5. Linear topological spaces. I. Title.
 QA322.C64 2015
 515'.73--dc23   2014048798

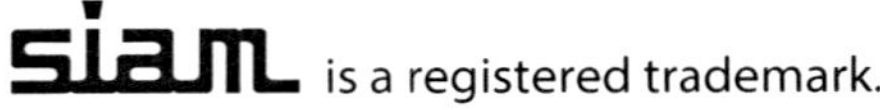

Contents

Preface

Parameter studies are everywhere in computational science. Complex engineering simulations must run several times with different inputs to effectively study the relationships between inputs and outputs. Studies like optimization, uncertainty quantification, and sensitivity analysis produce sophisticated characterizations of the input/output map. But thorough parameter studies are more difficult when each simulation is expensive and the number of parameters is large. In practice, the engineer may try to limit a study to the most important parameters, which effectively reduces the *dimension* of the parameter study.

Active subspaces offer a more general approach to reduce the study's dimension. They identify a set of important directions in the input space. If the engineer discovers a model's low-dimensional active subspace, she can exploit the reduced dimension to enable otherwise infeasible parameter studies for expensive simulations with many input parameters. This book develops active subspaces for dimension reduction in parameter studies.

My journey with active subspaces started at Stanford's Center for Turbulence Research Summer Program in 2010 with Gianluca Iaccarino (Stanford), Alireza Doostan (CU Boulder), and Qiqi Wang (MIT). We were studying surrogate models for calibrating the six inputs of an expensive scramjet simulation with Markov chain Monte Carlo. After struggling for two weeks to get the chain to converge, Qiqi came to the morning meeting touting some "magic parameters." He had found three orthogonal directions in the six-dimensional parameter space such that small input perturbations along these directions barely changed the pressure at prescribed sensor locations. The remaining three orthogonal directions gave three linear combinations of the six parameters; perturbing the latter linear combinations significantly changed the pressure, so they were more important for the calibration. He had found these directions by studying the pressure's gradient with respect to the input parameters. Lacking a technically descriptive name, Qiqi called the three important linear combinations *magic*. Given the scramjet runs we had available, he could build a regression surface in three variables with a higher degree polynomial than in six variables. We all thought this was a really cool idea.

The work was published in the Summer Program's proceedings and later as an AIAA conference paper [35]. But then it sat on the back burner for a while. In late 2011, I started bugging Qiqi that we should write a journal-worthy version of this *input dimension reduction* idea with more precise statements and formal development. Since so many uncertainty quantification techniques suffered from the dreaded curse of dimensionality, I thought reducing the dimension of a model's inputs could have a big impact. It took a while to get the formulation right, and we went back and forth with the reviewers. Amid that exchange, I found Trent Russi's 2010 Ph.D. thesis [108], in which he proposed a similar idea for reducing a model's input dimension with subspaces derived from random samples of the gradient; he called these subspaces *active subspaces*. We thought that was a catchy and reasonably descriptive name, so we adopted it.

The paper was finally accepted and appeared as [28]. Around the same time, Qiqi and I connected with Youssef Marzouk (MIT) and Tan Bui (UT Austin), who were working on techniques for improving Markov chain Monte Carlo for Bayesian inverse problems—including some subspace-based dimension reduction ideas. We put together a proposal for the DOE's Advanced Scientific Computing Research Applied Mathematics program, and we were awarded a three-year grant to develop active subspaces for data-intensive inverse problems. That grant has supported my effort to write this monograph.

Audience. Computational science is inherently interdisciplinary. I hope that this text offers something useful and interesting to researchers across fields. Dimension reduction must be practical and well-motivated—with easy-to-implement algorithms and easy- to-interpret results—to impact real applications. But the techniques must also be theoretically well-founded with rigorous performance guarantees for cases that satisfy simplifying assumptions. When simplified cases lack the challenges of real applications, engineers accuse the analysts of working on *toy problems*. When an engineering method has no rigorous performance guarantees, the analysts deride the method as *heuristic*—to which the engineer may respond that it works well in practice.

These squabbles are unlikely to subside in the near future, and this book will not resolve them. However, I have tried to balance both objectives. I have tried to develop the material so that it is amenable to analysis. And I have tried to demonstrate the value of active subspaces in engineering applications. I hope that both groups will benefit. There is still plenty to do on both fronts—analysis and engineering. Computational science graduate students may find a new perspective on their research with these techniques and subsequently advance the state of the art.

Outline. The first chapter provides a quick start with active subspaces for engineers paralyzed by their high-dimensional parameter studies. It offers some easy-to-implement procedures for discovering whether the given model admits an active subspace. If these tests show evidence of an active subspace, then the practitioner should be well-motivated to read on. The first chapter is reasonably self-contained. As such, I repeat some of the material in later chapters for further development.

The second chapter puts active subspaces into the broader context of algorithm research in uncertainty quantification. It includes important references to related statistics research on *sufficient dimension reduction* that blossomed more than 20 years ago.

The third and fourth chapters develop the technical content, including defining the active subspace, proposing and analyzing a method to discover the active subspace, and discussing strategies to exploit the active subspace for high-dimensional parameter studies.

The fifth chapter demonstrates the utility of active subspaces in three engineering applications: (i) studying the safe operating regime of a hypersonic scramjet, (ii) characterizing the relationship between model inputs and maximum power in a photovoltaic solar cell model, and (iii) optimizing the shape of an airfoil. Each section has just enough detail to put the application in the mathematical framework for active subspaces.

The quickest way to impact applications is to provide easy-to-use software. I have not included any language-specific implementations in the text—mostly because of the rapid pace of innovation in software and computing. However, I maintain the website `activesubspaces.org`, and I provide scripts and utilities there for working with active subspaces.

Acknowledgments. Throughout the text, I use the first person plural *we*[1] to reflect that the work in this book has been a collaborative effort. Although the choice and arrangement of words and symbols are my own, the intellectual content contains

---

[1]In the first chapter, I use the second person singular *you* to make directions easier to read.

significant contributions from my esteemed colleagues. My own interests cannot possibly cover the breadth of expertise needed for interdisciplinary computational science research—especially the domain expertise in the applications in Chapter 5. The final work would not have been possible without my colleagues' efforts.

I am very grateful for Qiqi Wang and Eric Dow at MIT. Much of the technical development in Chapters 3 and 4 comes from our paper [28]. David Gleich (Purdue) and I worked together to analyze the random sampling method for estimating active subspaces presented in Chapter 3, and his help with the presentation and several tricky turns in the proofs was invaluable [30]. Youssef Marzouk (MIT) and Tan Bui (UT Austin) helped refine my ideas on Bayesian inverse problems in Chapter 4. The scramjet application from Chapter 5 was a primary objective at Stanford's NNSA-funded Predictive Science Academic Alliance Program center. The specific work with active subspaces was performed jointly with Michael Emory (Stanford), Johan Larsson (UMD College Park), and Gianluca Iaccarino (Stanford) [29]. Their comments from the engineering side helped ensure that active subspaces could be valuable in real applications. The photovoltaics application in Chapter 5 consists of joint work with Brian Zaharatos (Colorado School of Mines) and Mark Campanelli (NREL) [37]. The airfoil shape optimization in Chapter 5 consists of joint work with Trent Lukaczyk, Francisco Palacios, and Juan Alonso at Stanford's Aerospace Design Laboratory [87]. I am especially thankful for Trent's help in running the simulations, preparing the figures, and interpreting the results in section 5.3. Lastly, I am grateful for the helpful suggestions and support from Ralph Smith (NCSU) and the *SIAM Spotlights* anonymous reviewers. Their comments helped me to clarify the presentation.

I have been extremely fortunate to receive funding from the U.S. Department of Energy. As a postdoc at Stanford, I was funded by the National Nuclear Security Administration under Award NA28614 through the Predictive Science Academic Alliance Program. As an assistant professor at Colorado School of Mines, my efforts have been supported by the Office of Science Advanced Scientific Computing Research Applied Mathematics program under Award DE-SC-0011077.

Chapter 1

Quick Start

We assume that you are here with a computer simulation of a complicated physical model that includes several input parameters. You want to perform some sort of parameter study—such as optimization, uncertainty quantification, or sensitivity analysis—with the simulation's predictions as a function of its inputs. But the methods you know for such studies are not practical with all of your parameters because of your limited computational budget. In other words, the methods need too much computation in *high dimensions*—more computation than you have available.

If you knew that some of the inputs would not affect the results of the parameter study, then you could ignore them. But we assume that you do not know this beforehand—at least not with certainty. One strategy to enable your parameter study would be to use some of the computational budget to discover which input parameters are important and which you can safely ignore. If you can ignore some inputs, then we say that you have *reduced the dimension* of the parameter study. In the best case, you reduce the dimension to the point that a thorough parameter study becomes feasible within your budget.

Active subspaces offer one appealing approach for this type of dimension reduction. However, instead of identifying a subset of the inputs as important, active subspaces identify a set of important directions in the space of all inputs. Each direction is a set of weights that define a linear combination of the inputs. If the simulation's prediction does not change as the inputs move along a particular direction, then we can safely ignore that direction in the parameter study. For example, Figure 1.1 plots the function $\exp(0.7x_1 + 0.3x_2)$. The arrows indicate the direction $[0.7, 0.3]$ along which this function varies the most and the orthogonal direction $[-0.3, 0.7]$ along which it is constant. This generalizes the idea of identifying a subset of important parameters, since a particular function could have important directions aligned with the input coordinates.

The rest of this chapter offers some easy-to-implement procedures for determining whether your simulation is eligible for dimension reduction with active subspaces. We assume you have the following pieces available:

1. a simulation model with m well-defined inputs and a scalar quantity of interest,

2. a range for each of the independent input parameters,

3. resources for running the simulation multiple times.

Some comments on these quantities are in order. First, the quantity of interest should be a smooth function of the input parameters. Generic discontinuities wreak havoc in the

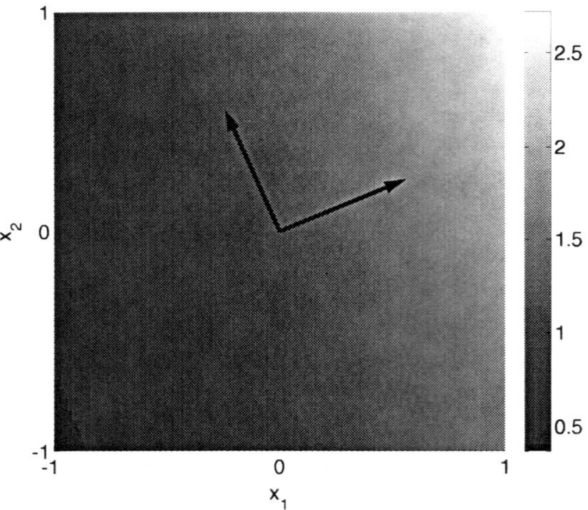

Figure 1.1. *The function* $\exp(0.7x_1 + 0.3x_2)$ *varies the most along the direction* $[0.7, 0.3]$ *and is constant along the orthogonal direction.*

Table 1.1. *Notation we use during the check for dimension reduction.*

Symbol	Meaning
$f : \mathbb{R}^m \to \mathbb{R}$	An abstract representation of the map from normalized inputs to the simulation's quantity of interest
$\mathbf{x}$	An m-vector containing the normalized input parameters of the simulation
$q = f(\mathbf{x})$	The scalar quantity of interest that depends on the inputs
$\nabla f(\mathbf{x})$	The gradient of the map

random sampling-based algorithms we use. Second, you may have some choice in defining the ranges of the input parameters. Make sure that your model is well-defined and numerically well-behaved for all parameter values within the ranges. To ensure this, you may need to adjust meshes or solver tolerances for different values of the inputs. Beware of asymptotes in your quantity of interest; unbounded growth can also cause problems in sampling-based approaches. Table 1.1 sets up working notation to describe the procedures, and we begin with a few preliminary preparations.

1. **Normalized inputs.** We need to normalize the input parameters to be centered at zero with equal ranges. Normalizing removes units and ensures that parameters with relatively large values do not disproportionately affect the analysis. Consider a vector $\mathbf{x}$ with m components, each between -1 and 1. If $\mathbf{x}_\ell$ and $\mathbf{x}_u$ are m-vectors containing the lower and upper bounds of the simulation's inputs, respectively, then the scaled and shifted vector

$$\frac{1}{2}\left(\operatorname{diag}(\mathbf{x}_u - \mathbf{x}_\ell)\mathbf{x} + (\mathbf{x}_u + \mathbf{x}_\ell)\right) \tag{1.1}$$

contains the natural inputs of the simulation.

2. **Sampling density.** The check for dimension reduction is based on randomly sampling the normalized inputs and running the model for each sample; random sampling is a powerful tool for working in high-dimensional spaces. If your model is not naturally endowed with such a density (most are not), then you must choose one. We suggest choosing something easy to work with, such as a uniform density over the hypercube $[-1, 1]^m$. We denote this density by ρ. The results depend on the choice of ρ, but that dependence may be weak. If you are worried by such dependence, you should repeat the analysis with different ρ's to see how the results change.

3. **The largest dimension.** Think ahead to how you plan to use the low-dimensional model. For example, if you ultimately want to build response surfaces, you may not be willing to work in more than three or four dimensions. Let k be one greater than the largest dimension that you can handle in subsequent use. In the worst case, you can choose $k = m + 1$, which is one more than the number of input parameters. However, you will save substantial computational effort if you choose k less than m.

4. **An oversampling factor.** Rules of thumb in random sampling algorithms include an oversampling factor. For example, it might be possible to fit a linear model with only $m + 1$ evaluations of the quantity of interest. But it is safer to fit the model with evaluations between $2m$ and $10m$ evaluations. Choose an oversampling factor α to be between 2 and 10.

With ρ, k, and α defined, we proceed to the computation.

1.1 ▪ Gradients or no gradients?

There are several ideas based on subspaces for reducing the dimension of the input space, many of which were developed in the context of regression models in statistics [38]. The active subspace in particular is derived from the gradient $\nabla f(\mathbf{x})$, and discovering an active subspace requires the capability to evaluate the gradient—or at least an approximate gradient—at any $\mathbf{x}$ in the space of input parameters. Fortunately, more and more simulation codes have gradient capabilities thanks to technologies such as adjoint solvers and algorithmic differentiation, although these are often absent in legacy codes and simulations that include multiple, coupled components.

If your simulation has gradient capabilities, start with Algorithm 1.1, which requires $M = \alpha k \log(m)$ evaluations of the gradient. (See Chapter 3 for how we derived this number.) This algorithm produces a set of eigenvectors and eigenvalues. We use the eigenvalues to determine the dimension of the active subspace, and the corresponding eigenvectors define the active subspace. If the number m of input parameters is in the tens to thousands, then this eigenvalue decomposition is quick to compute on a modern laptop.

If you do not have gradients, but your simulation is amenable to finite difference approximations of the gradient—i.e., numerical noise in your quantity of interest is small enough to use finite differences—then you can use Algorithm 1.1 with finite differences. First-order finite differences cost $m + 1$ simulation runs per gradient evaluation—one for the point in the input space where the approximate gradient is evaluated and one per perturbed parameter. The total cost of Algorithm 1.1 with first-order finite differences is $N = \alpha k(m + 1)\log(m)$ total simulations.

***ALGORITHM* 1.1. Active subspace estimation with gradients.**

1. Draw $M = \alpha k \log(m)$ independent samples $\{\mathbf{x}_i\}$ according to the sampling density ρ.

2. For each sample $\mathbf{x}_i$, compute the gradient $\nabla_{\mathbf{x}} f_i = \nabla f(\mathbf{x}_i)$ and the quantity of interest $q_i = f(\mathbf{x}_i)$.

3. Compute the matrix $\hat{C}$ and its eigenvalue decomposition,

$$\hat{C} = \frac{1}{M} \sum_{i=1}^{M} \nabla_{\mathbf{x}} f_i \nabla_{\mathbf{x}} f_i^T = \hat{W} \hat{\Lambda} \hat{W}^T, \tag{1.2}$$

where $\hat{W}$ is the matrix of eigenvectors, and $\hat{\Lambda} = \mathrm{diag}(\hat{\lambda}_1, \dots, \hat{\lambda}_m)$ is the diagonal matrix of eigenvalues ordered in decreasing order.

If you do not have access to gradients, and finite differences are infeasible, you can build a model of $f(\mathbf{x})$ to approximate gradients. For example, if you fit a polynomial model to $f(\mathbf{x})$, then computing gradients of the polynomial model is straightforward.[2] But fitting global multivariate polynomial models of degree greater than one requires many simulation runs. Kernel-based response surfaces, e.g., radial basis functions or Gaussian processes, suffer from the same scalability issue. Fortunately, we are primarily interested in *local* behavior when approximating gradients. We propose the following algorithm based on local linear models to estimate the eigenpairs from Algorithm 1.1 when gradients are not available. Each local linear model is fit with a subset of the predictions from a set of randomly sampled runs.

***ALGORITHM* 1.2. Active subspace estimation with local linear models.**

1. Choose $N \geq \alpha m$, $M = \alpha k \log(m)$, and an integer p such that $m + 1 \leq p \leq N$.

2. Draw N independent samples $\{\mathbf{x}_j\}$ according to the density ρ, and compute $q_j = f(\mathbf{x}_j)$ for each sample.

3. Draw M independent samples $\{\mathbf{x}_i'\}$ according to ρ.

4. For each $\mathbf{x}_i'$, find the p points from the set $\{\mathbf{x}_j\}$ nearest to $\mathbf{x}_i'$; denote this set by $\mathcal{X}_i$. Let $\mathcal{Q}_i$ be the subset of $\{q_j\}$ that corresponds to the points in $\mathcal{X}_i$.

5. Use least-squares to fit the coefficients c_i and $\mathbf{b}_i$ of a local linear regression model,

$$q_j \approx c_i + \mathbf{b}_i^T \mathbf{x}_j, \quad \mathbf{x}_j \in \mathcal{X}_i, \, q_j \in \mathcal{Q}_i. \tag{1.3}$$

6. Compute the matrix $\hat{C}$ and its eigenvalue decomposition,

$$\hat{C} = \frac{1}{M} \sum_{i=1}^{M} \mathbf{b}_i \mathbf{b}_i^T = \hat{W} \hat{\Lambda} \hat{W}^T. \tag{1.4}$$

---

[2]First-order finite differences are partial derivatives of local linear interpolants.

There is a lot to say about Algorithm 1.2. Note that it needs two additional parameters: (i) the number N of initial runs that are used to build the local linear models, and (ii) the parameter p that determines the number of runs used to fit each local linear model. We are actively analyzing Algorithm 1.2. We hope to derive quantitative statements about the quality of the approximated $\hat{W}$ and $\hat{\Lambda}$ along with more precise guidance for the choices of N and p. For now we are guided by intuition, a few observations, and our experience that this method has revealed active subspaces in real applications.

First, in step 2 we choose the points $\{\mathbf{x}_j\}$ independently at random according to ρ. There may be better choices for $\{\mathbf{x}_j\}$—e.g., quasi–Monte Carlo points [100]—that properly cover the space of inputs when $m < 10$. In general, the points $\{\mathbf{x}_j\}$ need to resolve the essential features of $f(\mathbf{x})$ for the local linear models to accurately approximate the gradients. Second, the algorithm assumes that each least-squares problem in step 5 is well-posed. If the matrix used to fit the linear model is rank deficient, then the fit needs either regularization or a larger value for p; including more neighbors when fitting the linear model may make the least-squares problem well-posed.

If $p = m + 1$, then each local linear model interpolates the predictions in $\mathcal{Q}_i$. If $p = N$, then there is only one global linear model, and $\hat{C}$ is rank one. The case when $p = N$ is sufficiently interesting to warrant its own algorithm, which reduces to fitting a global linear model and computing the normalized gradient. The global linear model is interesting for two reasons. First, it's so cheap! It only requires enough runs to fit a linear model. Second, we have used this approach to discover a one-dimensional active subspace in several real applications; see section 5.1 for an example.

ALGORITHM 1.3. Active subspace estimation with a global linear model.

1. Draw $N = \alpha m$ independent samples $\{\mathbf{x}_j\}$ from the sampling density ρ.

2. For each sample $\mathbf{x}_j$, run the simulation and compute the quantity of interest, $q_j = f(\mathbf{x}_j)$.

3. Use least-squares to compute the coefficients c and $\mathbf{b}$ of the linear model

$$q_j \approx c + \mathbf{b}^T \mathbf{x}_j, \quad j = 1, \dots, N. \tag{1.5}$$

4. Compute the normalized gradient of the linear model

$$\hat{\mathbf{w}} = \mathbf{b} / \|\mathbf{b}\|. \tag{1.6}$$

Algorithm 1.3 first fits a linear model of the quantity of interest as a function of the simulation's input parameters using $N = \alpha m$ simulation runs. The direction $\hat{\mathbf{w}}$ in (1.6) is the normalized gradient of the fitted linear model. If the map from inputs to quantity of interest is actually a linear function, then $\hat{\mathbf{w}}$ is equivalent to the first eigenvector from the gradient-based Algorithm 1.1; however, in general they are different. The global linear model can produce only a single direction.

Figure 1.2 compares the cost of each algorithm in terms of simulation runs as m increases. The comparison implicitly assumes that evaluating the gradient is about as expensive as running the simulation. The cost αm of the Algorithm 1.3, denoted "Linear" in the legend, is a lower bound for Algorithm 1.2. Next we describe how to use the quantities computed by the algorithms to decide if the model is eligible for dimension reduction.

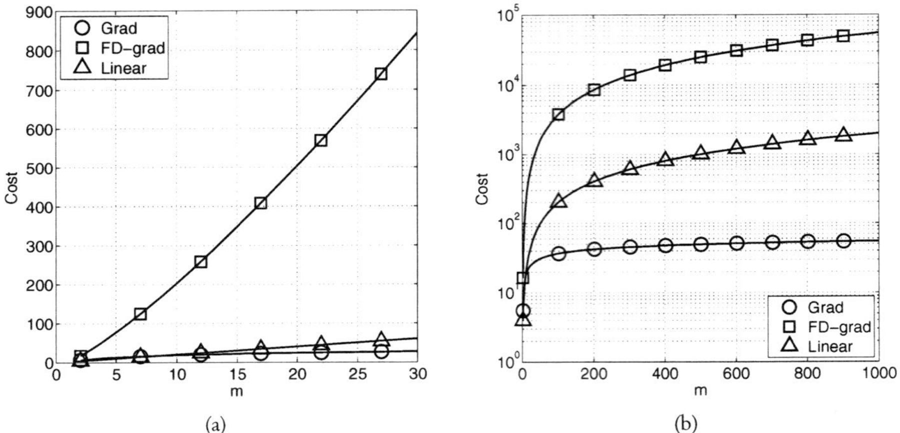

Figure 1.2. *The costs of Algorithm 1.1 (Grad), Algorithm 1.1 with first-order finite differences (FD-grad), and Algorithm 1.3 (Linear) in terms of number of function evaluations. These plots use $k = 4$ and $\alpha = 2$. The comparison assumes that evaluating the gradient is as expensive as running the simulation.* (a) *zooms in on m between 2 and 30.* (b) *shows m between 2 and 1000 on a log scale.*

1.2 ▪ Evaluating the dimension of the active subspace

We use the first k eigenvalues $\hat{\lambda}_1, \ldots, \hat{\lambda}_k$ from Algorithm 1.1 to check for an active subspace. Note that we listed only the first k eigenvalues instead of all m. The number $M = \alpha k \log(m)$ gives enough samples of the gradient to ensure that the first k eigenvalues are sufficiently accurate; see Chapter 3. Plot these eigenvalues—preferably on a log scale—and look for *gaps*. A gap in the eigenvalues indicates a separation between active and inactive subspaces. Additionally, the computed eigenvectors are more accurate when there is a gap in the eigenvalues. Choose the dimension n of the active subspace to be the number of eigenvalues preceding the gap. For example, if $k = 6$ and there is a gap between eigenvalues 3 and 4, then choose $n = 3$.

If there is no gap in the eigenvalues, then you do not have an active subspace up to dimension k. The map between your simulation's inputs and quantity of interest varies significantly along all of the directions that the first k eigenvectors represent. You might repeat the process with a larger k to look for gaps in later eigenvalues. Or you might increase α and sample more gradients to increase the accuracy of the eigenvalues. However, the slow convergence of the random sampling means that additional samples may not be of much help. In either case, if you decide to return for more samples of the gradient, you can reuse the ones you already computed since they are all independent samples.

Looking for gaps differs from other methods that use eigenvalues to produce a low-dimensional representation, such as the proper orthogonal decomposition for model reduction in dynamical systems or principal component analysis for high-dimensional data sets. For these analyses, one typically chooses the dimension of the low-dimensional approximation based on the magnitude of the eigenvalues—e.g., so that the sum of the retained eigenvalues exceeds some proportion of the sum of all the eigenvalues. We are less concerned with such energy-like criteria and more concerned with accurately approximating the subspace. The accuracy of the computed subspace depends heavily on the gap between associated eigenvalues; see Chapter 3 for the supporting analysis.

Gaps in the eigenvalues from Algorithm 1.1 with finite differences or Algorithm 1.2 with local linear models can similarly identify an active subspace. However, it is possible that the errors in approximate gradients can artificially increase an eigenvalue, apparently removing a gap. We show an example of this phenomenon in Chapter 3 on a simple test problem. Be sure to compare the value of your finite difference stepsize to the magnitude of the eigenvalues. As long as the eigenvalues are larger than the finite difference stepsize, any conclusions you draw from the eigenvalues are valid. If the eigenvalues are smaller than the finite difference stepsize, then you should decrease the stepsize for greater confidence.

Algorithm 1.3, which uses the global linear model, can produce only a one-dimensional subspace. If the dimension of the active subspace is greater than one, then Algorithm 1.3 does not discover the complete subspace. The power of the global linear model is apparent when used with the sufficient summary plots, which we define next.

1.3 • Sufficient summary plots

Sufficient summary plots were developed by Cook in the context of regression graphics [38]. They are powerful visualization tools for identifying low-dimensional structure in a quantity that depends on several input variables. At first glance, they are scatter plots. But closer inspection reveals great innovation in the horizontal axis. We introduce these plots using the function $f(x_1, x_2) = \exp(0.7x_1 + 0.3x_2)$ shown in Figure 1.1.

Most graphical software packages allow one to produce a surface plot of a function that depends on two inputs. The surface plots can be dragged and rotated to allow viewing the function from different angles. Imagine first dragging the plot so that the x–z plane is parallel to the plane of the computer monitor and the z-axis is aligned straight up and down. Next rotate the plot about the z-axis, which produces views of the surface from varying angles in the x–y plane. If one can find a view such that the surface collapses to a univariate plot, then the function f is constant along the direction into and out of the monitor. Such a collapse suggests that the two input variables can be condensed into a single variable, and f can be treated like a function of one variable instead of two. Voilà! Dimension reduction!

Figure 1.3(a) shows such a rotated view of $\exp(0.7x_1 + 0.3x_2)$. This is equivalent to a plot of $\exp(t)$ against the variable $t = 0.7x_1 + 0.3x_2$. In other words, it is a plot of the quantity of interest against a particular linear combination of the input variables.

The sufficient summary plot generalizes the idea of plotting the quantity of interest q against a linear combination of the input variables $\eta^T \mathbf{x}$, where η is an m-vector. We denote the sufficient summary plot by $\mathrm{SSP}(q, \eta^T \mathbf{x})$. If η is a canonical basis vector (a vector of zeros with a one in a single entry), then the sufficient summary plot becomes a simple scatter plot, such as those described in section 1.2.3 of Saltelli et al. [110]. The question is how to choose η such that a univariate trend emerges from the high-dimensional surface— that is, if one is present. Alternatively, can we find η_1 and η_2 to produce a surface of the quantity of interest as a function of the linear combinations $\eta_1^T \mathbf{x}$ and $\eta_2^T \mathbf{x}$? We denote the latter surface by $\mathrm{SSP}(q, \eta_1^T \mathbf{x}, \eta_2^T \mathbf{x})$.

The first two eigenvectors from $\hat{W}$ in (1.4) are natural choices for η_1 and η_2 in the sufficient summary plot. (Beware: If the inputs are not normalized to be centered at zero with equal ranges, then the sufficient summary plot may fail to reveal the dimension reduction.) More precisely, the one- and two-dimensional sufficient summary plots using the eigenvectors defining the active subspace are denoted $\mathrm{SSP}(q, \hat{\mathbf{w}}_1^T \mathbf{x})$ and $\mathrm{SSP}(q, \hat{\mathbf{w}}_1^T \mathbf{x}, \hat{\mathbf{w}}_2^T \mathbf{x})$, respectively. If a large gap exists between the first and second eigenvalues, then there is a good chance that a strong univariate trend is present in $\mathrm{SSP}(q, \hat{\mathbf{w}}_1^T \mathbf{x})$; similarly, a gap between the second and third eigenvalues suggests a strong bivariate trend. It is possible

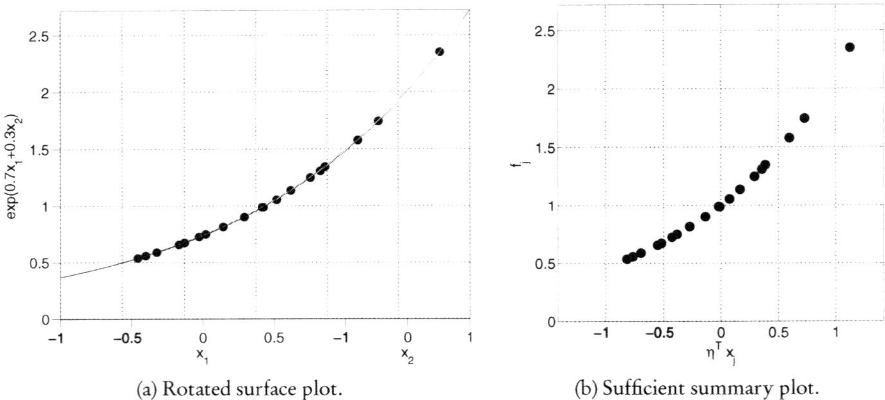

(a) Rotated surface plot.   (b) Sufficient summary plot.

Figure 1.3. (b) *is the sufficient summary plot using the direction* $\hat{\mathbf{w}}$ *produced by Algorithm 1.3 with the map* $f(x_1, x_2) = \exp(0.7x_1 + 0.3x_2)$. (a) *is a rotated surface plot of the same function.*

to construct a function that has an eigenvalue gap but lacks an apparent trend in the sufficient summary plot. But in all the models we have seen in practice, an eigenvalue gap corresponds to a sufficient summary plot that enables dimension reduction.

Roughly speaking, if there are high-frequency oscillations in the function and if the gradients resolve the oscillations, then the eigenvectors from Algorithm 1.1 may choose directions related to the oscillations instead of larger trends. The corresponding sufficient summary plot may not reveal reduced dimensionality, even if it is present with another choice of η. In this case, it may be better to choose the direction $\hat{\mathbf{w}}$ from Algorithm 1.3 or the eigenvectors from Algorithm 1.2 (with $p > m + 1$) than the eigenvectors from Algorithm 1.1, which uses the gradients. Since the linear models are fit with least-squares, they are naturally robust to noise; their surfaces are smooth approximations to the true function.

We presented the linear models as workarounds for when gradients are not available. However, if the quantity of interest is noisy—say, due to a finite number of fixed point iterations in the simulation's solver—then the least-squares-fit linear models provide complementary choices for η_1 and η_2. We can mimic such behavior with the function $\sin(4\pi x_1) + 4x_2$, where the sine term creates oscillations along x_1. Figure 1.4 shows the sufficient summary plots using Algorithms 1.1 and 1.3 with 208 samples of the gradient and the function, respectively. The sufficient summary plot using the direction (close to $[0, 1]$) produced by the linear model appears to identify a more useful dimension reduction than the one using the first eigenvector (close to $[1, 0]$) from the gradient-based algorithm.

There are cases when the directions computed by all algorithm variants coincide. The function $\exp(0.7x_1 + 0.3x_2)$ is such a case. Figure 1.3(b) shows the sufficient summary plot for this function using the direction $\eta = \hat{\mathbf{w}}$ computed with the linear model in Algorithm 1.3. This plot would be essentially the same if we had used the first eigenvector from the gradient-based approach (Algorithm 1.1) or the local linear model approach (Algorithm 1.2).

The global linear model has one pitfall worth mentioning. If the function is not monotonic—in other words, if the sign of some partial derivative changes over the input parameter space—then the linear model may fail to identify the dimension reduction space. We demonstrate this with the function $\frac{1}{2}(0.7x_1 + 0.3x_2)^2$. Figure 1.5 compares the

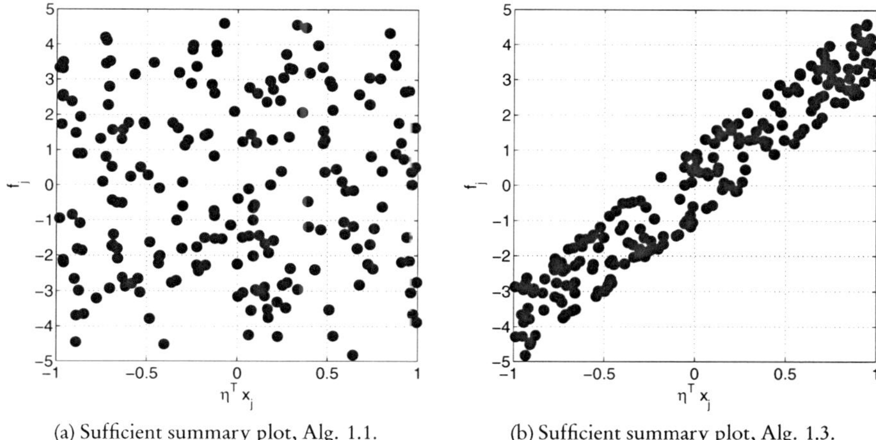

(a) Sufficient summary plot, Alg. 1.1. (b) Sufficient summary plot, Alg. 1.3.

Figure 1.4. (a) *is the sufficient summary plot using the first eigenvector produced by Algorithm 1.1 for the function* $f(x_1, x_2) = \sin(4\pi x_1) + 4x_2$. (b) *is the sufficient summary plot using the direction* $\hat{w}$ *computed with Algorithm 1.3. The directions produced by the least-squares linear models are robust to high-frequency oscillations.*

sufficient summary plots using the gradient-based eigenvector and the normalized gradient of the linear model; the plots use 208 samples of the gradient and the function, respectively. We have not encountered such cases in practice. We think this is because many—if not most—engineering quantities of interest are monotonic with respect to the inputs. Intuition-driven statements from engineers, such as *more of input X yields more/less of output Q within the regime of interest,* support this claim. In such cases, the linear model can often quickly and cheaply identify a one-dimensional subspace that is sufficient for describing the behavior of the quantity of interest. We present the results of Algorithm 1.3 applied to a simulation of a hypersonic scramjet in Chapter 5.

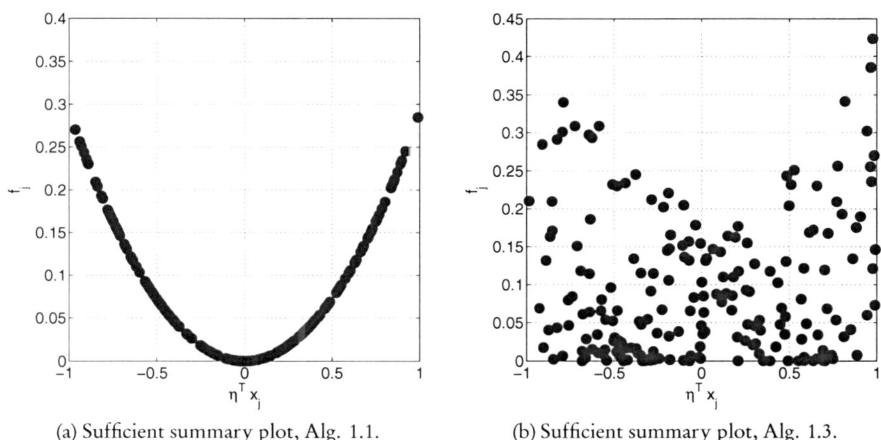

(a) Sufficient summary plot, Alg. 1.1. (b) Sufficient summary plot, Alg. 1.3.

Figure 1.5. (a) *is the sufficient summary plot using the first eigenvector produced by Algorithm 1.1 for the function* $f(x_1, x_2) = \frac{1}{2}(0.7x_1 + 0.3x_2)^2$. (b) *is the sufficient summary plot using the direction* $\hat{w}$ *computed with Algorithm 1.3. The linear model fails to identify the dimension reduction space when the function is not monotonic with respect to its inputs.*

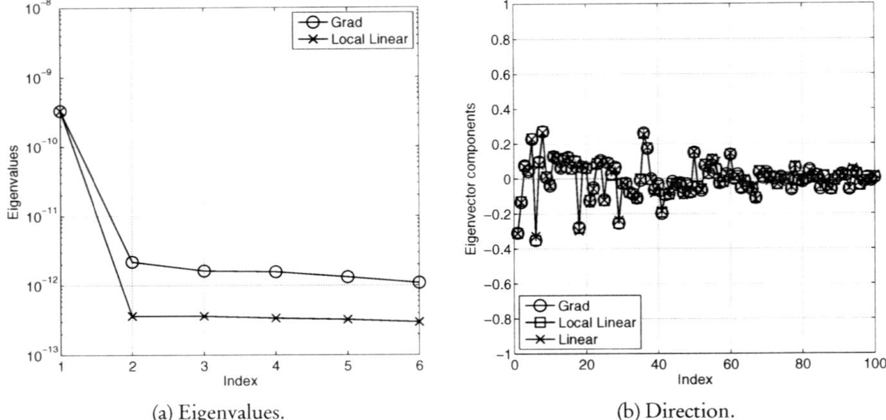

(a) Eigenvalues.   (b) Direction.

Figure 1.6. (a) *displays the eigenvalues computed with Algorithms* 1.1 *and* 1.2. (b) *shows the components of the first eigenvector from Algorithms* 1.1 *and* 1.2 *along with the components of the vector computed with the global linear model in Algorithm* 1.3. *In this case, all computed directions are reasonably consistent. Image* (b) *is modified from* [28].

1.4 ▪ An example with a parameterized PDE

We compare these approaches on Poisson's equation in two spatial dimensions with constant forcing and spatially varying coefficients parameterized by 100 Gaussian random variables. Similar models appear in single-phase subsurface flow, where the parameterized coefficients model uncertainty in subsurface permeability. More details of this specific test problem are given in section 3.4.2. The point of this example is to show the algorithms' results on something closer to an engineering application than simple test functions; see Chapter 5 for real engineering models.

We begin with the following three required pieces taken from the beginning of this chapter:

- The smooth scalar quantity of interest q is the average of the PDE solution on the right boundary.

- The $m = 100$ parameters that characterize the PDE coefficients are unbounded, but the model is well-defined for all parameter values.

- The simulation uses a finite element method to estimate the solution and the quantity of interest given values for the parameters. The computing platform is MATLAB 2014a on a MacBook Air with a dual-core processor and 8 GB of RAM.

We do not have a budget per se. The finite element simulation runs fast enough to permit us to run all algorithm variants. The sampling density ρ is a standard Gaussian density on $\mathbb{R}^{100}$; it is already normalized. We choose $k = 6$ for the largest dimension of interest and $\alpha = 3$ for an oversampling factor. Gradients are available via an adjoint solver.

We apply Algorithm 1.1 using the gradient computed with the adjoint solution. We use $M = 83$ evaluations of the gradient. Figure 1.6 shows the first $k = 6$ eigenvalues and the first eigenvector's components. Note the large gap between the first and second eigenvalues. Figure 1.6 also shows the eigenvalues and first eigenvector's components from Algorithm 1.2 with $N = 2000$ and $p = 180$. The eigenvalues have a similar gap between the first and the second, and the first eigenvector's components match the ones computed using gradients in Algorithm 1.1. We apply Algorithm 1.3 with $N = 300$; the components of its computed vector $\hat{\mathbf{w}}$ match the first eigenvector's components from Algorithms 1.1 and 1.2.

Figure 1.7 displays the sufficient summary plots from each of the three algorithm variants. All sufficient summary plots show a strong univariate trend in the function of 100 parameters, which is consistent with the eigenvalue analysis. Figure 1.7(a) plots the 83 evaluations q_i against $\hat{\mathbf{w}}_1^T \mathbf{x}_i$ from the gradient-based algorithm. Figure 1.7(b) shows 100 of the $N = 2000$ samples used in Algorithm 1.2. Figure 1.7(c) shows 100 of the 300 samples used to fit the global linear model in Algorithm 1.3.

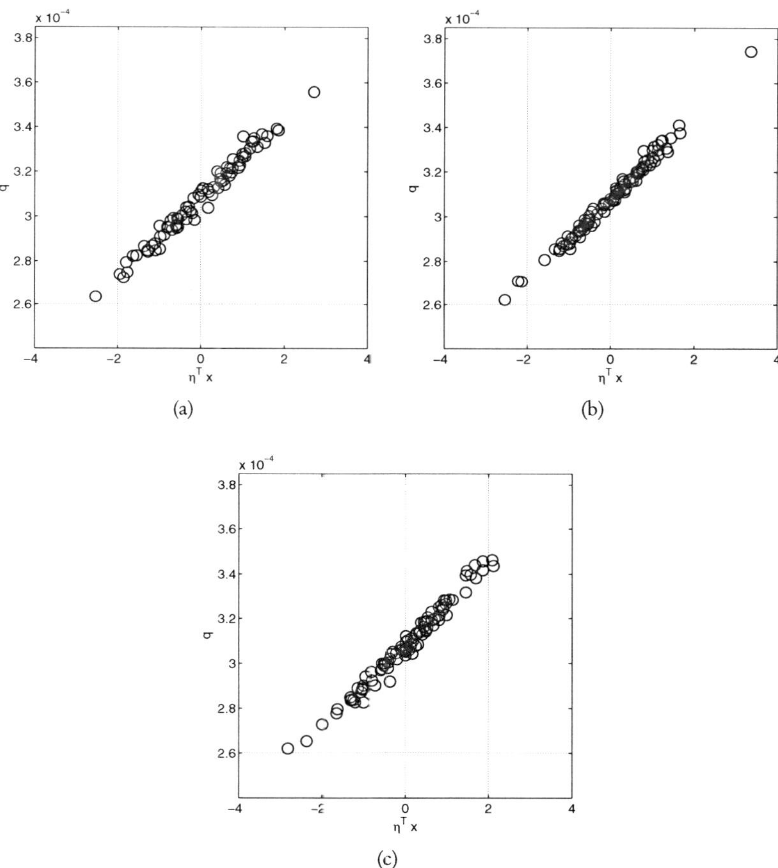

Figure 1.7. (a) *Sufficient summary plots from Algorithm 1.1 using 83 gradient samples.* (b) *Algorithm 1.2 using $N = 2000$ initial samples and $p = 180$.* (c) *Algorithm 1.3 using 300 samples. All plots indicate that a univariate model is a reasonable choice for this function. Image (a) is modified from* [28].

For this PDE model, the results from all algorithms are consistent, but this is not the case every time. You should choose the algorithm that is most appropriate for your particular application. The primary factors that influence this choice are (i) how many parameters your model has, (ii) how many times you can run the simulation, (iii) whether or not you have gradients, and (iv) whether or not there is noise in your quantity of interest.

Chapter 5 presents several real applications where we discuss how these considerations affect our choice of algorithm to search for dimension reduction.

One final note: The results of any random sampling algorithm change based on the particular set of samples that are drawn. In Chapter 3 we develop a bootstrap procedure for studying the variability in the estimated eigenvalues and active subspace. We can then include bootstrap intervals in the plots. The bootstrap is the most appropriate choice for studying variability because it is easy to implement and does not require further calls to the simulation code.

Chapter 2

Parameterized Models in Physics and Engineering

Today's models of our world are wonderfully complex thanks to fast and powerful computers. Modern simulations couple interacting physical phenomena to model rich, interconnected systems. Researchers on the frontier of computational science exploit high-performance computing to bridge the gap between micro- and macroscales. Next-generation simulations will combine physical models with high-resolution images and vast stores of measurement data to calibrate, validate, and improve predictions.

This complexity comes at a cost. Coupled systems are difficult to analyze. Predictions may oppose intuition, which leaves the scientist struggling to interpret and justify the results. Identifying a complex model's most important or influential components becomes more challenging as the number of components increases. If a complex model's prediction deviates from comparable measurements, the modelers wonder where they went wrong—and they shiver at the thought of exhaustively examining the potential deficiencies. In short, complexity breeds uncertainty.[3]

Uncertainty quantification (UQ) [112] has emerged in recent years offering tools to quantify sensitivity and variability in predictions from complex simulations. These tools combine methods from probability, statistics, approximation theory, optimization, and numerical analysis. The breadth of methods is rivaled only by the breadth of applications in need of UQ—each with its own notion of uncertainty—which makes it difficult to precisely and concisely define UQ.

We focus on a subclass of models that appears in UQ—namely *parameterized* simulations, where the predictions depend on some input parameters. The compelling inputs in parameterized simulations are those that affect the mathematical model underlying the simulation—as opposed to discretization parameters (e.g., grid spacing) or solver parameters (e.g., stopping tolerances). Ideally, simulation results have a well-defined relationship with discretization and solver parameters; greater resolution and smaller tolerances produce more accurate numerical approximations. Granted, this quixotic notion may not hold in practice; nevertheless, we assume away the troubles of insufficient numerical accuracy. And we attend to the relationship between the mathematical model's inputs—such as initial/boundary conditions, material properties, and external forcings—and the simulation's predictions. We denote this relationship by $f(\mathbf{x})$. Here $\mathbf{x}$ is a vector of the inputs and f maps those inputs to a prediction. The notation may raise specters of a much larger class of mathematical objects seen in analysis, and we try to address those concepts as needed. But in the back of our collective mind, $f(\mathbf{x})$ means a complex simulation's prediction given a set of inputs.

---

[3]That's ironic. Increasing complexity in a model is usually meant to reduce uncertainty.

Chapter 5 presents the following examples of parameterized simulations: a scramjet, where f is the pressure at the end of a combustor and $\mathbf{x}$ represents the operating conditions; a photovoltaic solar cell, where f is the maximum power and $\mathbf{x}$ represents circuit properties; and a transonic jet wing, where f is the wing's drag and $\mathbf{x}$ characterizes the wing's shape. Other examples include a self-igniting gas flame, where f is the ignition time and $\mathbf{x}$ represents the chemical reaction rates; an oil reservoir, where f is the amount of oil produced and $\mathbf{x}$ characterizes the permeability of the subsurface; the earth's climate, where f is the global average temperature and $\mathbf{x}$ represents the sources and sinks of CO_2; and a power grid, where f is the cost of keeping the lights on and $\mathbf{x}$ represents consumer demand.

For us, f is a single real number—a scalar *quantity of interest* derived from the simulation's results. Most complex simulations produce several numbers, and different quantities of interest are important to different researchers. We imagine that each researcher can treat her quantities of interest one at a time. The vector $\mathbf{x}$ contains several real numbers, and each real number is a *dimension* that can vary independently; we also use *dimension* to refer to the number of components in $\mathbf{x}$.

We assume that f changes continuously with changes in $\mathbf{x}$, i.e., that $f(\mathbf{x})$ is smooth. Nonsmooth or discontinuous f's can destroy the performance of computational methods that seek to characterize the relationship between $\mathbf{x}$ and f—especially when $\mathbf{x}$ has several components. Often, the quantity of interest may be smooth even when intermediate quantities are not. For example, pressure and velocity fields in compressible flow models may be discontinuous functions of $\mathbf{x}$, but derived quantities—such as the location of the shock or an integrated measure of pressure—may be continuous. If a modeler can choose between smooth and nonsmooth quantities of interest, we encourage her to choose the smooth ones.

Additionally, f must be differentiable with respect to $\mathbf{x}$, and we extensively use the gradient $\nabla f(\mathbf{x})$ to study $f(\mathbf{x})$. It is not unreasonable to assume that the simulation produces $\nabla f(\mathbf{x})$ along with $f(\mathbf{x})$. Techniques such as algorithmic differentiation [61] and adjoint methods [15, 21] are available in several simulation codes, which enable efficient optimization and control of quantities of interest. When gradients are not available, finite differences or related approximations may suffice. But beware of noise caused by numerical methods [91, 92].

2.1 ▪ What is random?

There is an ongoing discussion about the role of randomness in attempts to quantify uncertainty in a simulation's prediction [76]. The simulation is deterministic; a fixed value of $\mathbf{x}$ produces the same value of f each time the simulation runs. So what is random? Bayesian approaches treat f as an unknown function and model it with a Gaussian process, where a few runs at carefully selected input values inform the posterior density [68, 74]. Frequentist approaches employ design of experiments in the space of inputs to generate pairs $\{\mathbf{x}_i, f(\mathbf{x}_i)\}$, which are used to fit regression models [109]. In practice, these methods may produce the same results, although interpretations can differ.

While some have questioned the utility of probability for representing uncertainty [66], most prominent works in UQ algorithms treat $\mathbf{x}$ as a random vector with a given probability density function [58, 130, 79, 44, 52, 89, 31]. The density function on $\mathbf{x}$ is a necessary piece of information to ensure the well-posedness of the problem. If $\mathbf{x}$ is random, then $f(\mathbf{x})$ is a deterministic function of a random variable, which produces a random variable. Quantifying uncertainty is equated with computing statistics of f, such as its mean, variance, probability density function, or quantiles. When $\mathbf{x}$ is continuous with a known density and f is continuous, these computations reduce to integration. One can describe this integration with general measure theory concepts, but basic calculus typically suffices.

The trouble here is identifying the density function of $\mathbf{x}$. *Optimal uncertainty quantification* [101] computes a range of estimates for statistics over all possible density functions consistent with available information. Current research in UQ explores Bayesian methods for estimating the density function of $\mathbf{x}$ as a Bayesian posterior conditioned on given measurements corresponding to the simulation's output [118, 10], often dubbed the *inverse problem*. Alternative approaches to the inverse problem eschew the Bayesian perspective in favor of tools from analysis [13].

As we develop active subspaces, we need to integrate against the density function of $\mathbf{x}$, which we assume to be given. If many density functions are consistent with available information, then one should rightly analyze the sensitivity of the computed quantities with respect to changes in the density—but we do not pursue such ideas. Randomness enters our analysis and computations through randomly sampling $\mathbf{x}$ according to its given density function and running the simulation for each sample, i.e., Monte Carlo methods [100]. We use nonparametric bootstrap techniques [49] to study variability in estimates computed from samples.

2.2 ▪ The parameter studies

In practice, quantifying uncertainty in parameterized simulations often reduces to a question about the input parameters' effects on the predictions. The types of parameter studies are as varied as the types of simulations. To narrow our focus, we consider the following four specific types of parameter studies that arise frequently, each phrased as a question. We creatively refer to these questions as *the parameter studies*:

- [**Integration**] Given uncertainty in the inputs, what is the expected prediction, or what is the probability that it exceeds some threshold?

- [**Response surfaces**] Given a few simulation runs at carefully chosen values of $\mathbf{x}$, what is the model likely to predict at some other $\mathbf{x}$?

- [**Inversion**] What are the values of the inputs that bring the model into agreement with given measurements corresponding to the prediction?

- [**Optimization**] What is the range of predictions the model can produce given ranges of the input parameters?

Each question has a precise mathematical formulation, which we describe in Chapter 4. The answers may help quantify uncertainty—depending on how one defines uncertainty in a particular application—or they may provide more general insight into the model.

2.3 ▪ Too many parameters!

A limited computational budget, an expensive simulation, and more than a handful of parameters can paralyze a scientist attempting the parameter studies. How should she choose which runs to perform from among all the possible combinations of the inputs? Even with a given density for $\mathbf{x}$, a fully functioning simulation, and a precisely defined quantity of interest, performing any of the parameter studies may be intractable because of limited computational resources. If time and resources were unlimited, she could run the simulation at every possible floating point value of $\mathbf{x}$ and process the results to answer the parameter questions with great confidence. The trouble is how to answer them confidently with a limited number of runs.

There are two things fighting for the limited resources. First, each run might be expensive on its own, requiring extensive time on available computers. Several recent *model reduction* techniques exist that attempt to compute comparable predictions at reduced costs for the same inputs [18, 22, 24, 32, 34, 107]. However, using reduced order models in a parameter study produces answers for the reduced order model instead of the actual model.

The second problem is that the number of runs needed to complete a parameter study can be astronomically large when $\mathbf{x}$ has just a handful of components. For example, say that we need 10 runs per input parameter for a simulation with 10 input parameters to build an accurate response surface. That's 10^{10} runs! If each run completed in one second on one core and we had 10 cores available, it would take 32 years just to produce the data set. This is the *curse of dimensionality*, which loosely states that the number of evaluations needed for accurate approximation increases exponentially as the number of dimensions increases.

The only way to fight the curse of dimensionality is to identify and exploit special structure in the model. Global optimization may be feasible in several variables if f is a convex function [12]. Inversion may be possible in high dimensions if f is a linear function, the noise model on f is additive and Gaussian, and the prior density on $\mathbf{x}$ is Gaussian. Integration is possible if $f(\mathbf{x})$ is a polynomial of $\mathbf{x}$.

Several recent methods build response surfaces in high dimensions, assuming the map $f(\mathbf{x})$ admits special structure. If f can be reasonably well approximated by a sum of univariate functions,

$$f(\mathbf{x}) \approx f_1(x_1) + \cdots + f_m(x_m), \tag{2.1}$$

then sparse grid interpolation performs well [3, 129, 94, 33]. If f admits low-rank structure,

$$f(\mathbf{x}) \approx \sum_{i=1}^{r} f_{i,1}(x_1) \cdots f_{i,m}(x_m), \tag{2.2}$$

where r is much smaller than the number of parameters, then methods based on separation of variables perform well [9, 98, 5, 105, 45]. If f can be approximated by a sparse linear combination in some basis $\{\phi_i(\mathbf{x})\}$,

$$f(\mathbf{x}) \approx \sum_{i=1}^{n} a_i \, \phi_i(\mathbf{x}), \tag{2.3}$$

where many a_i are zero, then techniques for compressed sensing perform well [46, 104]. Unfortunately, it is hard to know beforehand whether a given $f(\mathbf{x})$ admits such exploitable structure, and identifying structure may require as much effort as any of the parameter studies.

2.4 ▪ Dimension reduction: Subsets and subspaces

Another approach would be to try to identify a subset of parameters that, when varied, do not cause significant change in the prediction. It is easy to exploit this type of structure—just ignore the input parameters that cause insignificant change and focus your effort on significant parameters. Limiting your attention from all inputs to a subset of important ones is a type of *dimension reduction*. Occasionally one can distinguish significant from insignificant parameters with off-the-cuff reasoning and back-of-the-envelope calculations. But a priori distinction is harder in more complex models. Instead, one may use a bit of the computational budget for preliminary experiments that discover which parameters are significant and which are not. The potential benefits of reducing the dimension may justify the cost of the discovery.

Sensitivity analysis ranks the input parameters in order of importance, but analyzing sensitivities is also a parameter study. The computation needed to confidently rank all of the input parameters may easily exceed the available resources in high dimensions, depending on chosen sensitivity measures. Local sensitivity measures perturb each input from a nominal value and measure the change in the prediction. These measures need one run per input beyond the nominal run—which is relatively cheap and often feasible in high dimensions. But local measures are fraught with difficulties. How large should the perturbation be? And what if perturbations from the nominal value do not represent perturbations at other points in the input space? Local measures explore the relationship between inputs and outputs in a very small part of the input space.

Measures of global sensitivity [110] address these difficulties by incorporating information from the entire input space. They are typically formulated as integrals over the parameter space, so they are costly to compute in high dimensions; see the first parameter study in section 2.2. Examples include quantities derived from a variance-based decomposition (e.g., functional ANOVA decomposition [113]) of $f(\mathbf{x})$ such as main effects, total effects, and related Sobol' indices [114, 99]. One can estimate these global sensitivity measures with Monte Carlo or quasi–Monte Carlo sampling. While slow to converge, Monte Carlo estimates of sensitivity measures may provide enough of a ranking to reduce the dimension. Alternatively, if one has an approximation of $f(\mathbf{x})$ as an orthogonal series, then estimates of Sobol' indices can be computed from the coefficients [119, 41]— although typically one must construct the orthogonal series approximation as a response surface. Other measures of global sensitivity are based on moments of partial derivatives of $f(\mathbf{x})$ [77] and the related *elementary effects* of Morris [93]. Some work has gone into studying the errors caused by neglecting dimensions with small sensitivity measures [115].

Global sensitivity measures identify subsets of input parameters that can be neglected to reduce the dimension. This type of dimension reduction is easy for the scientist to interpret, since each parameter relates to the science. Ignoring particular inputs means ignoring some relatively unimportant pieces of the model—where a computer experiment has discovered the relative unimportance. To exploit this dimension reduction, the scientist may attempt the four parameter studies by fixing the unimportant inputs at nominal values and varying only the remaining important inputs. But if she is willing to exert slightly more effort to interpret and exploit, then more powerful and more general tools for dimension reduction become available.

Instead of seeking an important subset of inputs, one may ask for important linear combinations of the inputs. This generalizes the subset-based dimension reduction; if all but one of the weights of the linear combination are all 0, then the linear combination identifies as important the one input parameter with the nonzero weight. The other extreme is a model where all parameters are important but only through a specific linear combination with roughly equal weights. Subset-based dimension reduction will not identify any possible dimension reduction, but the linear combination produces a single derived input that reduces the dimension to one. We call dimension reduction with linear combinations of inputs *subspace*-based dimension reduction.

Subspace-based dimension reduction is more general and more powerful than subset-based dimension reduction, but the power and generality come with two challenges. First, a linear combination of scientifically meaningful inputs is difficult to interpret. How does a climate scientist understand 2(cloud parameter) $+ 7$(vegetation parameter)? Proper scaling and normalization are essential for the weights to produce useful sensitivity measures on input parameters. The second challenge is how to exploit important linear combinations. When a subset of independent parameters is fixed, the remaining parameters may vary as they would otherwise. But fixing unimportant linear combinations and varying important ones requires care when attempting parameter studies—especially if the model's natural inputs $\mathbf{x}$ are bounded. We address both of these challenges in Chapter 4.

Regression models in statistics regularly exploit subspaces to reduce the dimension of predictors [38]. This idea is known as *sufficient dimension reduction* and is now an established research subfield in statistics. Techniques such as sliced inverse regression [81], sliced average variance estimation [40], principal Hessian directions [82], minimum average variance estimation [128], inverse regression estimation [39], contour regression [80], and sliced regression [123] seek weights such that the conditional density (or conditional mean) of the regression function, given a few linear combinations of predictors, is equal to the conditional density (or conditional mean) given all predictors. Some work exploits estimates of the regression function's gradient to identify the dimension reduction weights [69, 127, 54], but the gradient of the unknown regression function is generally considered too expensive to compute in high dimensions. Our computer simulation context differs from the regression context in three important respects. First, $f(\mathbf{x})$ is deterministic, so the conditional density of f given $\mathbf{x}$ is a delta function. Second, the simulation may compute the gradient $\nabla f(\mathbf{x})$ along with $f(\mathbf{x})$, and we use the gradient to construct the subspace used for dimension reduction. Third, the ultimate goal of the dimension reduction is not to compute the weights but to enable the parameter studies for simulations with several inputs.

Outside of statistics, subspaces are rarely used when approximating functions of several variables. Recently Fornasier, Schnass, and Vybiral [51] developed and analyzed a method based on compressed sensing to reconstruct functions of the form $f(\mathbf{x}) = g(A\mathbf{x})$, where A is a matrix with fewer rows than columns. Their technique uses only point evaluations of f, and it learns the elements of A from approximate directional derivatives.

In the context of computer simulations, Lieberman, Willcox, and Ghattas [84] describe a method for finding a subspace in a high-dimensional input space via a greedy optimization procedure. Russi's 2010 Ph.D. thesis [108] proposes a method that computes the singular value decomposition from a collection of randomly sampled gradients of the quantity of interest. Each singular vector contains the weights of a linear combination of the input parameters; the corresponding singular value quantifies the importance of that linear combination. He uses the important linear combinations to build quadratic approximations to $f(\mathbf{x})$. He defines the *active subspace* to be the space spanned by the important singular vectors. In practice, his construction of the linear combination weights is similar to ours, so we borrow his term. Mohammadi's related *sensitivity spaces* [90] are similarly derived from a collection of gradients of a quantity of interest. Stoyanov and Webster use the gradient to approximate a subspace for reducing the dimension of integrals of f coming from parameterized partial differential equations (PDEs) [117]. Tipireddy and Ghanem use subspaces derived from the coefficients of a polynomial approximation of $f(\mathbf{x})$ to reduce its dimension when computing statistics [120]. In aerospace design, Berguin and Mavris apply principal component analysis to a collection of gradients to find a low-dimensional subspace to exploit in design optimization [7, 6]. In nuclear engineering, Abdel-Khalik, Bang, and Wang employ subspaces within series approximations of quantities of interest to build reduced order models and surrogates for sensitivity analysis and uncertainty quantification [1]. Our previous work has applied active subspaces to design optimization in aerospace applications [25, 47], inverse analysis of hypersonic scramjet simulation [35], and spatial sensitivity analysis [36].

The active subspace is the span of the first few eigenvectors of a matrix derived from the gradient $\nabla f(\mathbf{x})$. In Chapter 3, we precisely define the active subspace and present examples to develop intuition about these eigenvectors. Loosely speaking, the eigenvectors define important directions in the input parameter space. Perturbing the inputs along the important directions causes greater change in the prediction, on average, than perturbing along the unimportant ones. We define the *active variables* to be linear combinations of the input parameters with weights from the important eigenvectors. The eigenvectors and eigenvalues are properties of the function and are not generally known. Chapter 3 develops and analyzes a Monte Carlo method for estimating the eigenvectors. It also considers the case when gradients are approximated, e.g., with finite differences.

Given estimates of the eigenvectors, we want to exploit any possible dimension reduction to enable otherwise intractable parameter studies. Chapter 4 devotes a section to each parameter study. We put the ideas to the test in Chapter 5 with a set of engineering applications.

Chapter 3

Discover the Active Subspace

Before we can discover the active subspace, we need to know what to look for. We first characterize a class of functions that covers many quantities of interest from parameterized simulations. We then define the active subspace, which is derived from a given function. Once we know what to look for, we propose and analyze a random sampling method to find it.

3.1 · Parameterized simulations and $f(\mathbf{x})$

Let $\mathbf{x}$ represent the input parameters to the simulation. The column vector $\mathbf{x}$ takes values in $\mathbb{R}^m$; we write $\mathbf{x} = [x_1, \ldots, x_m]^T$. We call m the *dimension* of the simulation. Do not confuse this dimension with other notions of dimension in the model, such as the number of spatial dimensions or the number of degrees of freedom in the discretized system. For us, dimension means the number of inputs to the model, where each input is a scalar. These scalars could parameterize a spatiotemporal field input, like the coefficients of a truncated Karhunen–Loève expansion of subsurface permeability or thermal conductivity. We assume that the modeler has already parameterized his or her continuous inputs with the m components of $\mathbf{x}$.

How large is m? In theory, m is larger than 1 and less than infinity. In practice, the methods we use to discover and exploit the active subspace are most appropriate when m is in the tens to thousands. This covers a large class of simulations. A method for parameter studies whose work scales exponentially with the dimension is not feasible for models with m in the tens to thousands. We limit m to thousands because we need to compute eigenvalue decompositions with $m \times m$ matrices; see section 3.3. When m is in the thousands, eigenvalue decompositions can be done quickly on a laptop. Really, the size of the eigenvalue problem is the limiting factor. If fancy, fast methods are available for large and dense eigenvalue problems, then m can be larger.

The space of inputs needs a positive weight function $\rho : \mathbb{R}^m \to \mathbb{R}$, or $\rho = \rho(\mathbf{x})$, because we need to compute integrals. We could use more general concepts from measure theory, but the added generality does not buy much in practice—and it might deter scientists from using active subspaces. If someone wants to make the analysis more general with measure theory, then she should go for it. (It might help take m to infinity.) We retain the current modest generality when we assume that the integral of ρ over $\mathbb{R}^m$ is 1, which makes it a probability density function. This brings the interpretation that $\mathbf{x}$ is a random vector, where the randomness models imprecisely specified input parameters. We also assume that ρ is bounded to avoid integrating unbounded functions. The mathematical quantities that define the active subspace are valid for any ρ that satisfies these conditions. However, the random sampling methods we use to discover the active subspace require something much stricter, namely, that we can draw independent samples from ρ.

We abstractly represent the map from simulation inputs to the quantity of interest by a function $f : \mathcal{X} \to \mathbb{R}$, where $\mathcal{X} \subseteq \mathbb{R}^m$ represents the set of interest in the input space. If $\mathcal{X}$ is bounded, then $\rho(\mathbf{x})$ is strictly positive for $\mathbf{x} \in \mathcal{X}$, and $\rho(\mathbf{x}) = 0$ for $\mathbf{x} \notin \mathcal{X}$. This focuses parameter studies on interesting parameter values. We assume that ρ and $\mathcal{X}$ are such that the components of $\mathbf{x}$ are independent with mean zero and similarly scaled—according to either the range or the variance. In practice, this means rotating, shifting, and/or scaling the input space appropriately. This normalization removes units and makes each input equally important for parameter studies. To keep the notation simple, we treat f as if it contains any necessary shifting and scaling to return a normalized $\mathbf{x}$ to the proper range of the simulation's inputs.

We assume that f is differentiable and Lipschitz continuous. The gradient is oriented as a column m-vector, and Lipschitz continuity implies that the gradient's norm is bounded, i.e.,

$$
\nabla_{\mathbf{x}} f(\mathbf{x}) = \begin{bmatrix} \frac{\partial f}{\partial x_1}(\mathbf{x}) \\ \vdots \\ \frac{\partial f}{\partial x_m}(\mathbf{x}) \end{bmatrix}, \qquad \|\nabla_{\mathbf{x}} f(\mathbf{x})\| \leq L, \quad \text{for all } \mathbf{x} \in \mathcal{X}, \tag{3.1}
$$

where $\| \cdot \|$ is the standard Euclidean norm. We assume that the simulation can produce the gradient of the quantity of interest, e.g., through adjoint methods, algorithmic differentiation, or some appropriate model such as finite differences.

In many cases, the modeler has some choice in ρ and $\mathcal{X}$. We recommend choosing something consistent with available information on the parameters that is also easy to work with, such as (i) a uniform distribution over a hyperrectangle built from bounds on each input parameter or (ii) a Gaussian distribution with variances chosen so that the bulk of the density function covers the inputs' ranges. These two choices are often used in practice. The active subspace quantities depend on the choice of ρ, so one should analyze the sensitivity of results with respect to the density.

3.2 ▪ Defining the active subspace

We begin with the following matrix, C, defined as the average of the outer product of the gradient with itself:

$$
C = \int (\nabla_{\mathbf{x}} f)(\nabla_{\mathbf{x}} f)^T \rho \, d\mathbf{x}. \tag{3.2}
$$

Each element of C is the average of the product of partial derivatives (which we assume exists),

$$
C_{ij} = \int \left(\frac{\partial f}{\partial x_i} \right) \left(\frac{\partial f}{\partial x_j} \right) \rho \, d\mathbf{x}, \qquad i, j = 1, \dots, m, \tag{3.3}
$$

where C_{ij} is the (i, j) element of C. Note that C is symmetric, and the size of C is $m \times m$, where m is the number of inputs to $f(\mathbf{x})$. If we consider $\nabla_{\mathbf{x}} f(\mathbf{x})$ to be a random vector by virtue of $\mathbf{x}$'s density ρ, then C is the uncentered covariance matrix of the gradient. Samarov studies this matrix as one of several *average derivative functionals* in the context of regression functions [111]. The diagonal elements of C are the mean-squared partial derivatives of f, which Kucherenko et al. use as sensitivity measures [77].

The matrix C is positive semidefinite, since

$$\mathbf{v}^T C \mathbf{v} = \int \left(\mathbf{v}^T(\nabla_{\mathbf{x}} f)\right)^2 \rho \, d\mathbf{x} \geq 0 \quad \text{for all } \mathbf{v} \in \mathbb{R}^m. \tag{3.4}$$

However, C is not derived from a Mercer kernel [106], so C is not a kernel matrix in the sense of kernel-based machine learning. Since C is symmetric, it has a real eigenvalue decomposition,

$$C = W \Lambda W^T, \qquad \Lambda = \text{diag}(\lambda_1, \ldots, \lambda_m), \qquad \lambda_1 \geq \cdots \geq \lambda_m \geq 0, \tag{3.5}$$

where W is the $m \times m$ orthogonal matrix whose columns $\{\mathbf{w}_1, \ldots, \mathbf{w}_m\}$ are the normalized eigenvectors of C. The following lemma quantifies the relationship between the gradient of f and the eigendecomposition of C.

Lemma 3.1. *The mean-squared directional derivative of f with respect to the eigenvector $\mathbf{w}_i$ is equal to the corresponding eigenvalue,*

$$\int \left((\nabla_{\mathbf{x}} f)^T \mathbf{w}_i\right)^2 \rho \, d\mathbf{x} = \lambda_i, \qquad i = 1, \ldots, m. \tag{3.6}$$

Proof. By the definition of C,

$$\lambda_i = \mathbf{w}_i^T C \mathbf{w}_i = \mathbf{w}_i^T \left(\int (\nabla_{\mathbf{x}} f)(\nabla_{\mathbf{x}} f)^T \rho \, d\mathbf{x}\right) \mathbf{w}_i = \int ((\nabla_{\mathbf{x}} f)^T \mathbf{w}_i)^2 \rho \, d\mathbf{x}, \tag{3.7}$$

as required.   □

Lemma 3.1 is worth a second look. The eigenvalues reveal very interesting properties of the function f. For example, if the smallest eigenvalue λ_m is exactly zero, then the mean-squared change in f along the eigenvector $\mathbf{w}_m$ is zero. Since $f(\mathbf{x})$ is continuous, this means the directional derivative $(\nabla_{\mathbf{x}} f)^T \mathbf{w}_m$ is zero everywhere in the domain $\mathcal{X}$. In other words, f is always constant along the direction defined by $\mathbf{w}_m$. Imagine standing at a point in the domain and looking at f along $\mathbf{w}_m$—it looks completely flat! Given a point $\mathbf{x}_1 \in \mathcal{X}$, choose a scalar z such that $\mathbf{x}_1 - z\mathbf{w}_m = \mathbf{x}_2 \in \mathcal{X}$. Then $f(\mathbf{x}_1) = f(\mathbf{x}_2)$, so f's value does not depend on z. This is information that we can exploit for dimension reduction.

The eigenvectors W define a rotation of $\mathbb{R}^m$ and consequently the domain of f. The arrows in Figure 1.1 show such a rotation in two dimensions. With the eigenvalues in decreasing order, we separate the rotated coordinates into two sets. On average, perturbations in the first set of coordinates change f more than perturbations in the second set of coordinates. We quantify the difference by separating the eigenvalues and eigenvectors,

$$\Lambda = \begin{bmatrix} \Lambda_1 & \\ & \Lambda_2 \end{bmatrix}, \qquad W = \begin{bmatrix} W_1 & W_2 \end{bmatrix}, \tag{3.8}$$

where $\Lambda_1 = \text{diag}(\lambda_1, \ldots, \lambda_n)$ with $n < m$, and W_1 contains the first n eigenvectors. Define the new variables $\mathbf{y}$ and $\mathbf{z}$ by

$$\mathbf{y} = W_1^T \mathbf{x} \in \mathbb{R}^n, \qquad \mathbf{z} = W_2^T \mathbf{x} \in \mathbb{R}^{m-n}. \tag{3.9}$$

Any $\mathbf{x} \in \mathbb{R}^m$ can be expressed in terms of $\mathbf{y}$ and $\mathbf{z}$,

$$\mathbf{x} = \underbrace{W W^T}_{I} \mathbf{x} = W_1 W_1^T \mathbf{x} + W_2 W_2^T \mathbf{x} = W_1 \mathbf{y} + W_2 \mathbf{z}. \tag{3.10}$$

Before going further, look at (3.10) once more. We exploit this decomposition several times in Chapter 4. In particular, $f(\mathbf{x}) = f(W_1 \mathbf{y} + W_2 \mathbf{z})$ implies that we can compute f's gradient with respect to $\mathbf{y}$ and $\mathbf{z}$. These partial derivatives are directional derivatives of f along the directions defined by the eigenvectors. The chain rule shows

$$\nabla_{\mathbf{y}} f(\mathbf{x}) = \nabla_{\mathbf{y}} f(W_1 \mathbf{y} + W_2 \mathbf{z}) = W_1^T \nabla_{\mathbf{x}} f(W_1 \mathbf{y} + W_2 \mathbf{z}) = W_1^T \nabla_{\mathbf{x}} f(\mathbf{x}). \tag{3.11}$$

Similarly, $\nabla_{\mathbf{z}} f(\mathbf{x}) = W_2^T \nabla_{\mathbf{x}} f(\mathbf{x})$. The next lemma relates the average inner product of the gradient with itself to the eigenvalues of C.

Lemma 3.2. *The mean-squared gradients of f with respect to $\mathbf{y}$ and $\mathbf{z}$ satisfy*

$$\int (\nabla_{\mathbf{y}} f)^T (\nabla_{\mathbf{y}} f) \rho \, d\mathbf{x} = \lambda_1 + \cdots + \lambda_n,$$
$$\int (\nabla_{\mathbf{z}} f)^T (\nabla_{\mathbf{z}} f) \rho \, d\mathbf{x} = \lambda_{n+1} + \cdots + \lambda_m. \tag{3.12}$$

Proof. Using the linearity of the trace,

$$\begin{aligned}
\int (\nabla_{\mathbf{y}} f)^T (\nabla_{\mathbf{y}} f) \rho \, d\mathbf{x} &= \int \mathrm{trace}\left((\nabla_{\mathbf{y}} f)(\nabla_{\mathbf{y}} f)^T \right) \rho \, d\mathbf{x} \\
&= \mathrm{trace}\left(\int (\nabla_{\mathbf{y}} f)(\nabla_{\mathbf{y}} f)^T \rho \, d\mathbf{x} \right) \\
&= \mathrm{trace}\left(W_1^T \left(\int (\nabla_{\mathbf{x}} f)(\nabla_{\mathbf{x}} f)^T \rho \, d\mathbf{x} \right) W_1 \right) \\
&= \mathrm{trace}\left(W_1^T C W_1 \right) \\
&= \mathrm{trace}\left(\Lambda_1 \right) \\
&= \lambda_1 + \cdots + \lambda_n,
\end{aligned} \tag{3.13}$$

as required. The derivation for the $\mathbf{z}$ components is similar. $\square$

The discussion that follows Lemma 3.1 applies here, too. Pretend that $\lambda_{n+1} = \cdots = \lambda_m = 0$. Then the gradient with respect to $\mathbf{z}$ is zero everywhere in the domain. Changing $\mathbf{z}$ does not change f at all. Imagine standing at a point in the domain and gazing at f. If one gazes along any direction in the range of the eigenvectors W_2, then f is flat!

There is a reason why scenic photos of the U.S. feature the Rocky Mountains more than the Great Plains. Flat landscapes are less interesting than changing landscapes. East-to-west road trippers exhale in relief upon seeing the mountains after hours of plains. Everyone tries to pass the flats as fast as possible. Analogously, we want to ignore directions along which f is constant and focus on directions in which it changes. This is the motivation behind the active subspace.

We define the *active subspace* to be the range of the eigenvectors in W_1. The *inactive subspace* is the range of the remaining eigenvectors in W_2. We call $\mathbf{y}$ the *active variables* and $\mathbf{z}$ the *inactive variables*. Recall the decomposition $\mathbf{x} = W_1 \mathbf{y} + W_2 \mathbf{z}$, so that

$$f(\mathbf{x}) = f(W_1 \mathbf{y} + W_2 \mathbf{z}). \tag{3.14}$$

Lemma 3.2 tells us that on average, minuscule perturbations in the active variables $\mathbf{y}$ change f more than minuscule perturbations in the inactive variables $\mathbf{z}$. How much more is quantified by the eigenvalues of C. The stipulation "on average" is very important. There may be specific sets of points in the domain where $\mathbf{z}$ perturbations change f more than $\mathbf{y}$ perturbations. But these sets are relatively small.

Two clarifying remarks are in order to address potential confusion. First, the active subspace is not a subset of the input space. We are not necessarily restricting our attention to input parameters in the range of the eigenvectors W_1. That would greatly limit our study of the relationship between inputs and outputs. We explore the interaction between the domain and the active subspace in Chapter 4. Second, we are not particularly interested in a low-rank approximation,

$$C \approx W_1 \Lambda_1 W_1^T. \tag{3.15}$$

Such approximations and their properties are classical. Instead, we want to approximate f by a function of fewer than m variables to enable parameter studies.

3.2.1 ▪ Examples

We examine two classes of functions to gain insight into the active subspace. The first class of functions is index models [83]. Let A be a full-rank matrix with m rows and $k < m$ columns. A k-index model has the form $f(\mathbf{x}) = h(A^T\mathbf{x})$, where $h : \mathbb{R}^k \to \mathbb{R}$. By the chain rule,

$$\nabla_{\mathbf{x}} f(\mathbf{x}) = A \nabla h(A^T\mathbf{x}), \tag{3.16}$$

where ∇h is the vector of partial derivatives of h with respect to its k arguments. Then C from (3.2) becomes

$$C = A \left(\int \nabla h(A^T\mathbf{x}) \big(\nabla h(A^T\mathbf{x}) \big)^T \rho \, d\mathbf{x} \right) A^T. \tag{3.17}$$

In this case, the eigenvalues $\lambda_{k+1} = \cdots = \lambda_m = 0$, so the rank of C is at most k. If we choose $n = k$ in the partition (3.8) and if the integrated matrix in the middle of (3.17) has full rank, then $\operatorname{ran}(W_1) = \operatorname{ran}(A)$. The notation $\operatorname{ran}(\cdot)$ with a matrix argument is shorthand for the range of the columns of the matrix. In the special case where $k = 1$, the one-dimensional active subspace can be found with a single evaluation of the gradient, as long as the derivative of h is not zero. In recent work [51], Fornasier, Schnass, and Vybiral develop random sampling methods to approximate index models using only point evaluations of f.

The second class contains functions of the form $f(\mathbf{x}) = h(\mathbf{x}^T B\mathbf{x})/2$, where $h : \mathbb{R} \to \mathbb{R}$, and B is a symmetric $m \times m$ matrix. In this case

$$C = B \left(\int \big(h'(\mathbf{x}^T B\mathbf{x}) \big)^2 \mathbf{x}\mathbf{x}^T \rho \, d\mathbf{x} \right) B, \tag{3.18}$$

where h' is the derivative of h. This implies that the null space of C is the null space of B provided that h' is nondegenerate. We study the example where $h(s) = s$ in section 3.4.1.

3.3 ▪ Computing the active subspace

The eigenpairs of C are properties of f, like the Lipschitz constant or Fourier coefficients. Each function's eigenvalues are different. Two different quantities of interest from the same simulation may have different eigenvalues. In a few special cases, we can make statements about the eigenpairs by examining f, but in general we need to compute them. The results in this section can be found in [30].

The tremendous potential benefits of dimension reduction motivate us to study methods for estimating the eigenpairs of C. But the elements of C are m-dimensional integrals, so computing W and Λ requires high-dimensional integration. Deterministic numerical integration rules are impractical beyond a handful of variables, especially if evaluating the integrand is costly. We focus on a random sampling approach to approximate the eigenpairs of C, where we take advantage of recent theoretical results that estimate the number of samples needed to approximate the spectrum of sums of random matrices.

If drawing independent samples from the density ρ is cheap and simple, then Algorithm 3.1 presents a straightforward and easy-to-implement random sampling method to estimate the eigenvalues Λ and eigenvectors W.

ALGORITHM 3.1. Random sampling to estimate the active subspace.

1. Draw M samples $\{\mathbf{x}_j\}$ independently according to the density function ρ.

2. For each $\mathbf{x}_j$, compute $\nabla_{\mathbf{x}} f_j = \nabla_{\mathbf{x}} f(\mathbf{x}_j)$.

3. Approximate

$$C \approx \hat{C} = \frac{1}{M} \sum_{j=1}^{M} (\nabla_{\mathbf{x}} f_j)(\nabla_{\mathbf{x}} f_j)^T. \tag{3.19}$$

4. Compute the eigendecomposition $\hat{C} = \hat{W} \hat{\Lambda} \hat{W}^T$.

The last step is equivalent to computing the singular value decomposition (SVD) of the matrix

$$\frac{1}{\sqrt{M}} \begin{bmatrix} \nabla_{\mathbf{x}} f_1 & \cdots & \nabla_{\mathbf{x}} f_M \end{bmatrix} = \hat{W} \sqrt{\hat{\Lambda}} \hat{V}, \tag{3.20}$$

where standard manipulations show that the singular values are the square roots of the eigenvalues, and the left singular vectors are the eigenvectors. The SVD perspective was developed by Russi [108] as the method for discovering the active subspace. It shows that the active subspace is related to the principal components of a collection of gradients.

For many simulations, the number m of input parameters is small enough (e.g., tens to thousands) so that computing the full eigendecomposition or SVD is negligible compared to the cost of computing the gradient M times; this is our case of interest. Therefore, we want to know how large M must be so that the estimates $\hat{\Lambda}$ and $\hat{W}$ are close to the true Λ and W.

To answer these questions, we apply recent work by Tropp [121] and Gittens and Tropp [59] on the spectrum of sums of random matrices. The gradient vector $\nabla_{\mathbf{x}} f(\mathbf{x})$ is a deterministic function of $\mathbf{x}$, so what is random? Randomness comes from the independent samples of $\mathbf{x}$ from the density ρ—a standard interpretation of Monte Carlo integration.

We use the following notation to develop the results. For two symmetric matrices A and B, $A \preceq B$ means that $B - A$ is positive semidefinite. The notation $\lambda_k(\cdot)$ and $\lambda_{\max}(\cdot)$ with matrix arguments denote the kth eigenvalue and the maximum eigenvalue of the matrix, respectively. All vector norms are the Euclidean norm, and all matrix norms are the norm induced by the Euclidean norm (i.e., the spectral norm).

Theorem 3.3. *Assume that $\|\nabla_\mathbf{x} f\| \leq L$ for all $\mathbf{x} \in \mathscr{X}$. Then for $\varepsilon \in (0,1]$,*

$$\mathbb{P}\left[\hat{\lambda}_k \geq (1+\varepsilon)\lambda_k\right] \leq (m-k+1)\exp\left(\frac{-M\lambda_k\varepsilon^2}{4L^2}\right) \tag{3.21}$$

and

$$\mathbb{P}\left[\hat{\lambda}_k \leq (1-\varepsilon)\lambda_k\right] \leq k\exp\left(\frac{-M\lambda_k^2\varepsilon^2}{4\lambda_1 L^2}\right). \tag{3.22}$$

The key to establishing Theorem 3.3 is a matrix Bernstein inequality from Theorem 5.3 of Gittens and Tropp [59]. When we apply this concentration result, we set

$$X_j = \nabla_\mathbf{x} f_j \nabla_\mathbf{x} f_j^T, \tag{3.23}$$

so that X_j is a random draw of the outer product of the gradient with itself. Thus, each X_j is an independent random sample from the same matrix-valued distribution. Under this notion of randomness,

$$\mathbb{E}\left[X_j\right] = \int \nabla_\mathbf{x} f_j \nabla_\mathbf{x} f_j^T \, \rho(\mathbf{x}_j)\, d\mathbf{x}_j = \int \nabla_\mathbf{x} f \nabla_\mathbf{x} f^T \, \rho \, d\mathbf{x} = C. \tag{3.24}$$

For completeness, we restate Theorem 5.3 from [59].

Theorem 3.4 (eigenvalue Bernstein inequality for subexponential matrices [59, Thm. 5.3]). *Consider a finite sequence $\{X_j\}$ of independent, random, symmetric matrices with dimension m, all of which satisfy the subexponential moment growth condition*

$$\mathbb{E}\left[X_j^p\right] \preceq \frac{p!}{2}B^{p-2}\Sigma_j^2 \quad \text{for } p = 2,3,4,\ldots,$$

where B is a positive constant and Σ_j^2 are positive-semidefinite matrices. Given an integer $k \leq m$, set

$$\mu_k = \lambda_k\left(\sum_j \mathbb{E}\left[X_j\right]\right).$$

Choose V_+ as an orthogonal matrix of size $m \times m - k + 1$ that satisfies

$$\mu_k = \lambda_{\max}\left(\sum_j V_+^T(\mathbb{E}X_j)V_+\right),$$

and define

$$\sigma_k^2 = \lambda_{\max}\left(\sum_j V_+^T \Sigma_j^2 V_+\right).$$

Then for any $t \geq 0$,

$$\mathbb{P}\left[\lambda_k\left(\sum_j X_j\right) \geq \mu_k + t\right] \leq \begin{cases} (m-k+1)\cdot\exp\{-\tfrac{1}{4}t^2/\sigma_k^2\}, & t \leq \sigma_k^2/B, \\ (m-k+1)\cdot\exp\{-\tfrac{1}{4}t/B\}, & t \geq \sigma_k^2/B. \end{cases}$$

Proof. *(Theorem* 3.3.*)* We begin with the upper estimate (3.21). First, note that

$$\mathbb{P}\left[\lambda_k(\hat{C}) \geq \lambda_k(C) + t\right] = \mathbb{P}\left[\lambda_k\left(\sum_{j=1}^{M}\nabla_{\mathbf{x}}f_j\nabla_{\mathbf{x}}f_j^{T}\right) \geq M\lambda_k + Mt\right]. \tag{3.25}$$

In this form we can apply Theorem 3.4. We check that the bound on the gradient's norm implies that the matrix $\nabla_{\mathbf{x}}f\nabla_{\mathbf{x}}f^{T}$ satisfies the subexponential growth condition:

$$\int \left(\nabla_{\mathbf{x}}f\nabla_{\mathbf{x}}f^{T}\right)^{p}\rho\,d\mathbf{x} = \int \left(\nabla_{\mathbf{x}}f^{T}\nabla_{\mathbf{x}}f\right)^{p-1}\nabla_{\mathbf{x}}f\nabla_{\mathbf{x}}f^{T}\rho\,d\mathbf{x}$$

$$\preceq \left(L^2\right)^{p-1}\int \nabla_{\mathbf{x}}f\nabla_{\mathbf{x}}f^{T}\rho\,d\mathbf{x} \tag{3.26}$$

$$\preceq \frac{p!}{2}\left(L^2\right)^{p-2}\left(L^2 C\right).$$

Next we set

$$\mu_k = \lambda_k\left(\sum_{j=1}^{M}\int \nabla_{\mathbf{x}}f\nabla_{\mathbf{x}}f^{T}\rho\,d\mathbf{x}\right) = M\lambda_k, \tag{3.27}$$

where we simplified using the identically distributed samples of $\mathbf{x}_j$. Choose $\boldsymbol{W}_{+} = \boldsymbol{W}(:,k:m)$ to be the last $m-k+1$ eigenvectors of C, and note that

$$\lambda_{\max}\left(\sum_{j=1}^{M}\boldsymbol{W}_{+}^{T}\left(\int \nabla_{\mathbf{x}}f\nabla_{\mathbf{x}}f^{T}\rho\,d\mathbf{x}\right)\boldsymbol{W}_{+}\right) = M\lambda_{\max}(\boldsymbol{W}_{+}^{T}C\boldsymbol{W}_{+}) = M\lambda_k = \mu_k,$$

$$\tag{3.28}$$

as required by Theorem 3.4. Define

$$\sigma_k^2 = \lambda_{\max}\left(\sum_{j=1}^{M}\boldsymbol{W}_{+}^{T}(L^2 C)\boldsymbol{W}_{+}\right) = ML^2\lambda_{\max}\left(\boldsymbol{W}_{+}^{T}C\boldsymbol{W}_{+}\right) = ML^2\lambda_k. \tag{3.29}$$

With these quantities, Theorem 3.4 states

$$\mathbb{P}\left[\lambda_k\left(\sum_{j=1}^{M}\nabla_{\mathbf{x}}f_j\nabla_{\mathbf{x}}f_j^{T}\right) \geq M\lambda_k + Mt\right] \leq (m-k+1)\exp\left(\frac{-(Mt)^2}{4\sigma_k^2}\right) \tag{3.30}$$

when $Mt \leq \sigma_k^2/L^2$. Applying this theorem with $t = \varepsilon\lambda_k(C)$ and the computed quantities produces the upper estimate (3.21).

For the lower estimate,

$$\mathbb{P}\left[\lambda_k(\hat{C}) \leq \lambda_k(C) - t\right]$$

$$= \mathbb{P}\left[-\lambda_k(\hat{C}) \geq -\lambda_k(C) + t\right]$$

$$= \mathbb{P}\left[-\lambda_k\left(\sum_{j=1}^{M}\nabla_{\mathbf{x}}f_j\nabla_{\mathbf{x}}f_j^{T}\right) \geq -M\lambda_k(C) + Mt\right]$$

$$= \mathbb{P}\left[\lambda_{m-k+1}\left(\sum_{j=1}^{M}\left(-\nabla_{\mathbf{x}}f_j\nabla_{\mathbf{x}}f_j^{T}\right)\right) \geq M\lambda_{m-k+1}(-C) + Mt\right] \tag{3.31}$$

$$= \mathbb{P}\left[\lambda_{k'}\left(\sum_{j=1}^{M}\left(-\nabla_{\mathbf{x}}f_j\nabla_{\mathbf{x}}f_j^{T}\right)\right) \geq M\lambda_{k'}(-C) + Mt\right]$$

for $k' = m - k + 1$. We can now apply Theorem 3.4 again. The subexponential moment growth condition is satisfied since

$$\int \left(-\nabla_{\mathbf{x}} f \nabla_{\mathbf{x}} f^T \right)^p \rho \, d\mathbf{x} \preceq \int \left(\nabla_{\mathbf{x}} f \nabla_{\mathbf{x}} f^T \right)^p \rho \, d\mathbf{x} \preceq \frac{p!}{2} \left(L^2 \right)^{p-2} \left(L^2 C \right). \tag{3.32}$$

Set

$$\mu_{k'} = \lambda_{k'} \left(\sum_{j=1}^{M} \int \left(-\nabla_{\mathbf{x}} f \nabla_{\mathbf{x}} f^T \right) \rho \, d\mathbf{x} \right) = M \lambda_{k'}(-C). \tag{3.33}$$

Set $W_+ = W(:, 1:k)$ to be the first k eigenvectors of C, and note that

$$\lambda_{\max} \left(\sum_{j=1}^{M} W_+^T \left(\int \left(-\nabla_{\mathbf{x}} f \nabla_{\mathbf{x}} f^T \right) \rho \, d\mathbf{x} \right) W_+ \right) = M \lambda_{\max} \left(-W_+^T C W_+ \right)$$

$$\begin{aligned}
&= M(-\lambda_k(C)) \\
&= M \lambda_{m-k+1}(-C) \\
&= M \lambda_{k'}(-C),
\end{aligned} \tag{3.34}$$

as required by Theorem 3.4. Set

$$\sigma_{k'}^2 = \lambda_{\max} \left(\sum_{j=1}^{M} W_+^T (L^2 C) W_+ \right) = M L^2 \lambda_{\max} \left(W_+^T C W_+ \right) = M L^2 \lambda_1. \tag{3.35}$$

Theorem 3.4 states

$$\mathbb{P} \left[\lambda_{k'} \left(\sum_{j=1}^{M} \left(-\nabla_{\mathbf{x}} f_j \nabla_{\mathbf{x}} f_j^T \right) \right) \geq M \lambda_{k'}(-C) + Mt \right] \leq k \exp \left(\frac{-(Mt)^2}{4\sigma_{k'}^2} \right) \tag{3.36}$$

when $Mt \leq \sigma_{k'}^2 / L^2$. Plug in the computed quantities with $t = -\varepsilon \lambda_{k'}(-C) = \varepsilon \lambda_k(C)$ to achieve the lower estimate (3.22). Note that the conditions on the estimate are that $\varepsilon \leq 1$. $\square$

We use this result to derive a lower bound on the number of gradient samples needed for relative accuracy of ε. Recall that the *big omega* notation $a = \Omega(b)$ means that a is bounded below by a constant times b.

Corollary 3.5. *Let $\varkappa_k = \lambda_1 / \lambda_k$. Then for $\varepsilon \in (0, 1]$,*

$$M = \Omega \left(\frac{L^2 \varkappa_k^2}{\lambda_1 \varepsilon^2} \log(m) \right) \tag{3.37}$$

implies $|\hat{\lambda}_k - \lambda_k| \leq \varepsilon \lambda_k$ with high probability.

Proof. Starting with the upper estimate from Theorem 3.3, if

$$M \geq \frac{4L^2}{\lambda_k \varepsilon^2} (\beta + 1) \log(m) \geq \frac{4L^2}{\lambda_k \varepsilon^2} (\beta \log(m) + \log(m - k + 1)), \tag{3.38}$$

then

$$\mathbb{P} \left[\hat{\lambda}_k \geq (1 + \varepsilon) \lambda_k \right] \leq m^{-\beta}. \tag{3.39}$$

Similarly for the lower estimate from Theorem 3.3, if

$$M \geq \frac{4L^2\lambda_1}{\lambda_k^2\varepsilon^2}(\beta+1)\log(m) \geq \frac{4L^2\lambda_1}{\lambda_k^2\varepsilon^2}(\beta\log(m)+\log(k)), \tag{3.40}$$

then

$$\mathbb{P}\left[\hat{\lambda}_k \leq (1-\varepsilon)\lambda_k\right] \leq m^{-\beta}. \tag{3.41}$$

Setting $\varkappa_k = \lambda_1/\lambda_k$ and taking

$$M \geq (\beta+1)\frac{4L^2\varkappa_k^2}{\lambda_1\varepsilon^2}\log(m) \tag{3.42}$$

satisfies both conditions, as required.   □

The quantities L, $\varkappa$, λ_1, and λ_k are properties of the given function f, although they are generally not known for f's derived from complex simulations. While m is also a property of f, the logarithmic scaling with the dimension is appealing. We exploit this for a general heuristic in section 3.4.

We can combine standard results from Golub and Van Loan [60, Chapter 8] on stability of invariant subspaces with results from Tropp [121] to obtain an estimate of the distance between the true active subspace and the estimated active subspace defined by $\hat{W}_1$. This requires a different matrix Bernstein inequality, now in the form of Theorem 6.1 from Tropp [121]. This theorem is restated below; we apply it with $X_j = \nabla_{\mathbf{x}}f_j\nabla_{\mathbf{x}}f_j^T - C$. In other words, the random matrix samples are the deviance of the jth sampled gradient outer product from the true matrix C.

Theorem 3.6 (matrix Bernstein: bounded case [121, Thm. 6.1]). *Consider a finite sequence $\{X_j\}$ of independent, random, symmetric matrices with dimension m. Assume that*

$$\mathbb{E}\left[X_j\right]=0 \quad \text{and} \quad \lambda_{\max}(X_j) \leq R \quad \text{almost surely.}$$

Compute the norm of the total variance,

$$\sigma^2 := \left\|\sum_j \mathbb{E}\left[X_j^2\right]\right\|.$$

Then the following inequality holds for all $t \geq 0$:

$$\mathbb{P}\left[\lambda_{\max}\left(\sum_j X_j\right) \geq t\right] \leq \begin{cases} m\exp(-3t^2/8\sigma^2), & t \leq \sigma^2/R, \\ m\exp(-3t/8R), & t \geq \sigma^2/R. \end{cases}$$

Theorem 3.7. *Assume $\|\nabla_{\mathbf{x}}f\| \leq L$ for all $\mathbf{x} \in \mathscr{X}$. Then for $\varepsilon \in (0,1]$,*

$$\mathbb{P}\left[\|\hat{C}-C\| \geq \varepsilon\|C\|\right] \leq 2m\exp\left(\frac{-3M\lambda_1\varepsilon^2}{8L^2}\right). \tag{3.43}$$

Proof. Observe that

$$\mathbb{P}\left[\|\hat{C}-C\|\geq t\right]=\mathbb{P}\left[\lambda_{\max}(\hat{C}-C)\geq t \text{ or } \lambda_{\max}(C-\hat{C})\geq t\right]$$

$$\leq \mathbb{P}\left[\lambda_{\max}(\hat{C}-C)\geq t\right]+\mathbb{P}\left[\lambda_{\max}(C-\hat{C})\geq t\right]$$

$$= \mathbb{P}\left[\lambda_{\max}\left(\sum_{j=1}^{M}(\nabla_{\mathbf{x}}f_j\nabla_{\mathbf{x}}f_j^T-C)\right)\geq Mt\right] \tag{3.44}$$

$$+\mathbb{P}\left[\lambda_{\max}\left(\sum_{j=1}^{M}(C-\nabla_{\mathbf{x}}f_j\nabla_{\mathbf{x}}f_j^T)\right)\geq Mt\right]$$

$$\leq 2\theta,$$

where θ upper-bounds both probabilities. The final result of the proof is the upper bound θ. Note that

$$\int (\nabla_{\mathbf{x}}f\nabla_{\mathbf{x}}f^T-C)\rho\,d\mathbf{x} = \int (C-\nabla_{\mathbf{x}}f\nabla_{\mathbf{x}}f^T)\rho\,d\mathbf{x} = 0. \tag{3.45}$$

Now, C being positive semidefinite and $\|\nabla_{\mathbf{x}}f\|\leq L$ imply

$$\lambda_{\max}(\nabla_{\mathbf{x}}f\nabla_{\mathbf{x}}f^T-C)= \max_{\|\mathbf{v}\|=1}\ \mathbf{v}^T\left(\nabla_{\mathbf{x}}f\nabla_{\mathbf{x}}f^T-C\right)\mathbf{v}$$

$$\leq \max_{\|\mathbf{v}\|=1}\ \mathbf{v}^T\left(\nabla_{\mathbf{x}}f\nabla_{\mathbf{x}}f^T\right)\mathbf{v} \leq L^2, \tag{3.46}$$

and this also holds for $\lambda_{\max}(C-\nabla_{\mathbf{x}}f\nabla_{\mathbf{x}}f^T)$, giving us the upper bound R for Theorem 3.6. We can bound the variance parameter σ^2 as

$$\sigma^2 = \left\|\left(\sum_{j=1}^{M}\int (\nabla_{\mathbf{x}}f_j\nabla_{\mathbf{x}}f_j^T-C)^2\rho\,d\mathbf{x}\right)\right\|$$

$$= M\left\|\int (\nabla_{\mathbf{x}}f\nabla_{\mathbf{x}}f^T-C)^2\rho\,d\mathbf{x}\right\| \tag{3.47}$$

$$\leq M\left\|L^2C-C^2\right\| \leq M\|C\|\left\|L^2I-C\right\| \leq M\lambda_1 L^2.$$

The last line follows from the fact that $\lambda_1 \leq L^2$. Again, this bound holds for $C-\nabla_{\mathbf{x}}f\nabla_{\mathbf{x}}f^T$. Theorem 3.6 holds for an upper bound on σ^2, which yields an upper bound on θ. Plugging in the computed quantities with $t = \varepsilon\|C\|=\varepsilon\lambda_1$ yields the desired result. $\square$

We use this result to produce a lower bound on the number of samples needed for ε relative accuracy.

Corollary 3.8. *For $\varepsilon \in (0, 1]$,*

$$M = \Omega\left(\frac{L^2}{\lambda_1\varepsilon^2}\log(m)\right) \tag{3.48}$$

implies that $\|\hat{C}-C\|\leq \varepsilon\|C\|$ with high probability.

Again, this bound depends on L and λ_1, which are properties of the given f. To control the error in the estimated subspace, we can combine Corollary 3.8 with standard error estimates for approximate invariant subspaces from numerical analysis. We quantify this error by the distance between the subspace defined by the range of W_1 and the subspace defined by the range of $\hat{W}_1$. Recall the definition of the distance between subspaces [116],

$$\text{dist}(\text{ran}(W_1), \text{ran}(\hat{W}_1)) = \|W_1 W_1^T - \hat{W}_1 \hat{W}_1^T\| = \|W_1^T \hat{W}_2\|. \tag{3.49}$$

In particular, we use Corollary 8.1.11 from Golub and Van Loan [60]; for convenience, we restate this corollary as a lemma in the form we need.

Lemma 3.9 (from [60, Cor. 8.1.11]). *Let C and $\hat{C} = C + E$ be symmetric $m \times m$ matrices with respective eigenvalues $\lambda_1, \ldots, \lambda_m$ and $\hat{\lambda}_1, \ldots, \hat{\lambda}_m$ and eigenvector matrices*

$$W = \begin{bmatrix} W_1 & W_2 \end{bmatrix}, \quad \hat{W} = \begin{bmatrix} \hat{W}_1 & \hat{W}_2 \end{bmatrix},$$

where W_1 and $\hat{W}_1$ contain the first $n < m$ columns. If $\lambda_n > \lambda_{n+1}$ and

$$\|E\| \leq \frac{\lambda_n - \lambda_{n+1}}{5},$$

then

$$\text{dist}(\text{ran}(W_1), \text{ran}(\hat{W}_1)) \leq \frac{4\|W_2^T E W_1\|}{\lambda_n - \lambda_{n+1}}.$$

Corollary 3.10. *Let $\varepsilon > 0$ be such that*

$$\varepsilon \leq \frac{\lambda_n - \lambda_{n+1}}{5\lambda_1}, \tag{3.50}$$

and choose M according to Corollary 3.8. Then with high probability,

$$\text{dist}(\text{ran}(W_1), \text{ran}(\hat{W}_1)) \leq \frac{4\lambda_1 \varepsilon}{\lambda_n - \lambda_{n+1}}. \tag{3.51}$$

Proof. Let $E = \hat{C} - C$ as in Lemma 3.9. For ε in (3.50) with M chosen according to Corollary 3.8, we have with high probability

$$\|E\| \leq \varepsilon \|C\| = \varepsilon \lambda_1 \leq (\lambda_n - \lambda_{n+1})/5, \tag{3.52}$$

which satisfies the conditions of Lemma 3.9. Then

$$\text{dist}(\text{ran}(W_1), \text{ran}(\hat{W}_1)) \leq 4\|E\|/(\lambda_n - \lambda_{n+1}) \leq 4\lambda_1 \varepsilon /(\lambda_n - \lambda_{n+1}), \tag{3.53}$$

as required. $\square$

In fact, the precise statement in Golub and Van Loan's Corollary 8.1.11 shows something much stronger using the matrix $W_2^T E W_1$. There is probably room to improve the bound in Corollary 3.10. Nevertheless, this corollary shows that control of the eigenvalues implies control of the subspace generated by the eigenvectors.

The error in the estimated subspace is inversely proportional to the corresponding gap in the eigenvalues. This implies, for example, that if the gap between the second and third eigenvalues is larger than the gap between the first and second, then estimates of a two-dimensional active subspace are more accurate than estimates of a one-dimensional active subspace. We show an example of this phenomenon in section 3.4.1.

3.3.1 ▪ Approximate gradients

Large-scale simulations often lack simple-to-evaluate expressions for the gradient of the quantity of interest with respect to the inputs. Finite difference approximations are useful when m is not too large and f is neither too expensive nor too noisy. Moré and Wild's recent work characterizes the gradient when function evaluations contain noise [91, 92].

Next we extend the bounds on errors in the estimated eigenpairs to the case when the gradient is computed with some error. The error model we analyze is generic. Let $\mathbf{g}(\mathbf{x})$ denote the approximate gradient computed at $\mathbf{x} \in \mathcal{X}$. We assume that

$$\|\mathbf{g}(\mathbf{x}) - \nabla_{\mathbf{x}} f(\mathbf{x})\| \leq \sqrt{m}\gamma_h, \quad \mathbf{x} \in \mathcal{X}, \tag{3.54}$$

where

$$\lim_{h \to 0} \gamma_h = 0. \tag{3.55}$$

The parameter h may be, for example, a finite difference parameter or grid spacing in a discrete adjoint computation.

Define the symmetric positive-semidefinite matrix G and its eigenvalue decomposition as

$$G = \int \mathbf{g}\mathbf{g}^T \rho \, d\mathbf{x} = U\Theta U^T, \quad \Theta = \mathrm{diag}(\theta_1, \dots, \theta_m), \tag{3.56}$$

and denote its random sample approximation as

$$\hat{G} = \frac{1}{M}\sum_{j=1}^{M} \mathbf{g}_j\,\mathbf{g}_j^T = \hat{U}\hat{\Theta}\hat{U}^T, \quad \hat{\Theta} = \mathrm{diag}(\hat{\theta}_1, \dots, \hat{\theta}_m), \tag{3.57}$$

where $\mathbf{g}_j = \mathbf{g}(\mathbf{x}_j)$ for $\mathbf{x}_j$ drawn independently according to ρ. With these quantities defined, we have the following lemma.

Lemma 3.11. *The norm of the difference between $\hat{C}$ and $\hat{G}$ is bounded by*

$$\|\hat{C} - \hat{G}\| \leq (\sqrt{m}\gamma_h + 2L)\sqrt{m}\gamma_h. \tag{3.58}$$

Proof. Let $\mathbf{g} = \mathbf{g}(\mathbf{x})$ and $\nabla_{\mathbf{x}} f = \nabla_{\mathbf{x}} f(\mathbf{x})$. First, observe

$$\|\mathbf{g} + \nabla_{\mathbf{x}} f\| = \|\mathbf{g} - \nabla_{\mathbf{x}} f + 2\nabla_{\mathbf{x}} f\| \leq \|\mathbf{g} - \nabla_{\mathbf{x}} f\| + 2\|\nabla_{\mathbf{x}} f\| \leq \sqrt{m}\gamma_h + 2L. \tag{3.59}$$

Next,

$$\begin{aligned}
\|\mathbf{g}\mathbf{g}^T - \nabla_{\mathbf{x}} f \nabla_{\mathbf{x}} f^T\| &= \frac{1}{2}\|(\mathbf{g} + \nabla_{\mathbf{x}} f)(\mathbf{g} - \nabla_{\mathbf{x}} f)^T + (\mathbf{g} - \nabla_{\mathbf{x}} f)(\mathbf{g} + \nabla_{\mathbf{x}} f)^T\| \\
&\leq \|(\mathbf{g} + \nabla_{\mathbf{x}} f)(\mathbf{g} - \nabla_{\mathbf{x}} f)^T\| \\
&\leq (\sqrt{m}\gamma_h + 2L)\sqrt{m}\gamma_h.
\end{aligned} \tag{3.60}$$

Then

$$\begin{aligned}
\|\hat{G} - \hat{C}\| &= \left\| \frac{1}{M}\sum_{j=1}^{M} \mathbf{g}_j\mathbf{g}_j^T - \frac{1}{M}\sum_{j=1}^{M} \nabla_{\mathbf{x}} f_j \nabla_{\mathbf{x}} f_j^T \right\| \\
&\leq \frac{1}{M}\sum_{j=1}^{M} \|\mathbf{g}_j\mathbf{g}_j^T - \nabla_{\mathbf{x}} f_j \nabla_{\mathbf{x}} f_j^T\| \\
&\leq \sqrt{m}\gamma_h(\sqrt{m}\gamma_h + 2L). \qquad \square
\end{aligned} \tag{3.61}$$

We combine Lemma 3.11 with Corollary 3.5 to study the error in the eigenvalues of the random sample estimate with approximate gradients.

Theorem 3.12. *For $\varepsilon \in (0,1]$, if M is chosen as (3.37), then the difference between the true eigenvalue λ_k and the eigenvalue $\hat{\theta}_k$ of the random sample estimate with approximate gradients is bounded as*

$$|\lambda_k - \hat{\theta}_k| \leq \varepsilon \lambda_k + \sqrt{m}\gamma_h(\sqrt{m}\gamma_h + 2L), \tag{3.62}$$

with high probability.

Proof. Observe that

$$|\lambda_k - \hat{\theta}_k| \leq |\lambda_k - \hat{\lambda}_k| + |\hat{\lambda}_k - \hat{\theta}_k|. \tag{3.63}$$

Apply Corollary 3.5 to the first term. The second term follows from [60, Corollary 8.1.6] combined with Lemma 3.11, since

$$|\hat{\theta}_k - \hat{\lambda}_k| = |\lambda_k(\hat{G}) - \lambda_k(\hat{C})| \leq \|\hat{G} - \hat{C}\| \leq \sqrt{m}\gamma_h(\sqrt{m}\gamma_h + 2L). \qquad \square \tag{3.64}$$

There are two sources of error in the bound from Theorem 3.12: finite sampling error and gradient approximation error. The gradient approximation error in the eigenvalue estimates goes to zero at the same rate as the error in the approximate gradients. Next we attend to the error in the active subspace computed with random samples and approximate gradients.

Theorem 3.13. *Choose ε according to Corollary 3.10, and choose M to satisfy both Corollary 3.10 and Corollary 3.5 for the $n + 1$th eigenvalue. Choose h small enough so that*

$$\sqrt{m}\gamma_h(\sqrt{m}\gamma_h + 2L) \leq \frac{\hat{\lambda}_n - \hat{\lambda}_{n+1}}{5}. \tag{3.65}$$

Then the distance between the true active subspace and the estimated active subspace with finite samples and approximate gradients is bounded by

$$\mathrm{dist}(\mathrm{ran}(\hat{U}_1), \mathrm{ran}(W_1)) \leq \frac{4\sqrt{m}\gamma_h(\sqrt{m}\gamma_h + 2L)}{(1-\varepsilon)\lambda_n - (1+\varepsilon)\lambda_{n+1}} + \frac{4\lambda_1\varepsilon}{\lambda_n - \lambda_{n+1}}, \tag{3.66}$$

with high probability.

Proof. First,

$$\mathrm{dist}(\mathrm{ran}(\hat{U}_1), \mathrm{ran}(W_1)) \leq \mathrm{dist}(\mathrm{ran}(\hat{U}_1), \mathrm{ran}(\hat{W}_1)) + \mathrm{dist}(\mathrm{ran}(\hat{W}_1), \mathrm{ran}(W_1)). \tag{3.67}$$

The second term is bounded in Corollary 3.10 under the assumptions on M and ε. By the condition (3.65) on h, Lemma 3.9 implies that the first term is bounded by

$$\mathrm{dist}(\mathrm{ran}(\hat{U}_1), \mathrm{ran}(\hat{W}_1)) \leq \frac{4\|\hat{G} - \hat{C}\|}{\hat{\lambda}_n - \hat{\lambda}_{n+1}}. \tag{3.68}$$

When M is large enough to satisfy Corollary 3.5, $|\hat{\lambda}_n - \lambda_n| \le \varepsilon \lambda_n$ and $|\hat{\lambda}_{n+1} - \lambda_{n+1}| \le \varepsilon \lambda_{n+1}$ with high probability. Then

$$
\begin{aligned}
\lambda_n - \lambda_{n+1} &= |\lambda_n - \lambda_{n+1}| \\
&\le |\lambda_n - \hat{\lambda}_n| + |\hat{\lambda}_{n+1} - \lambda_{n+1}| + (\hat{\lambda}_n - \hat{\lambda}_{n+1}) \\
&\le \varepsilon \lambda_n + \varepsilon \lambda_{n+1} + (\hat{\lambda}_n - \hat{\lambda}_{n+1}).
\end{aligned}
\tag{3.69}
$$

Rearranging this inequality yields

$$
\hat{\lambda}_n - \hat{\lambda}_{n+1} \ge (1-\varepsilon)\lambda_n - (1+\varepsilon)\lambda_{n+1}.
\tag{3.70}
$$

Combining this with the bound from Lemma 3.11 yields the result.   □

In summary, the eigenvalues and the eigenvectors approximated with random sampling and approximate gradients are well-behaved. The error bounds include a term that goes to zero like the error in the approximate gradient and a term that behaves like the random sample approximation with exact gradients.

3.4 ▪ A practical recipe

The bounds we present in section 3.3 provide a solid theoretical foundation for understanding the behavior of the random sampling estimates. However, many of the quantities in the bounds may not be known a priori, such as the maximum gradient norm L and the true eigenvalues of the matrix C. In this section we offer a practical recipe guided by the insights from the theory.

We estimate the eigenvalues with Algorithm 3.1. To approximate the first k eigenvalues of C with dimension $m \times m$, we recommend choosing the number M of independent gradient samples as

$$
M = \alpha k \log(m).
\tag{3.71}
$$

The constant α is an oversampling factor, which we typically choose to be between 2 and 10. This is a fudge factor for the unknown quantities in the theoretical bounds. It is likely that the specific quantities from the bounds are larger than 10. However, the Bernstein-type inequalities used to derive the bounds are likely conservative—and are probably loose for simulation models in practice. We have found this range of α to be sufficient in the applications we have studied. The $\log(m)$ term follows from the bounds in Theorem 3.3. We have plotted this heuristic for M as a function of m with $k = 6$ in Figure 3.1.

We suggest using a bootstrap to assess the variability in both the eigenvalue estimates and the subspace estimates. Efron and Tibshirani use the bootstrap to get empirical density functions of estimated eigenvalues and eigenvector components in section 7.2 of their well-known book [49]. Chapter 3 of Jolliffe's book [71] also comments on the bootstrap approach for estimating eigenvalues and eigenvectors of a covariance matrix from independent samples. The bootstrap estimates of the standard error and confidence intervals for the eigenvalues may be biased, but this bias decreases as the number M of samples increases. Since these estimates may be biased, we refer to them as *bootstrap intervals* instead of *confidence intervals*. Algorithm 3.2 describes the bootstrap procedure given samples of the gradient.

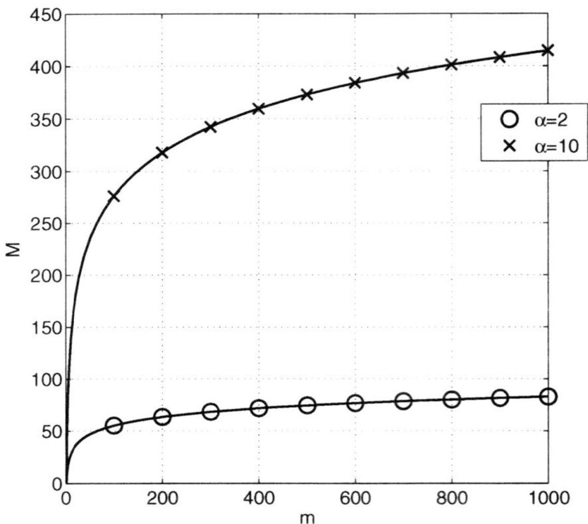

Figure 3.1. *The scaling with dimension m of a heuristic for the number M of gradient samples from (3.71) for $k = 6$. The lower and upper bounds use $\alpha = 2$ and $\alpha = 10$, respectively.*

ALGORITHM 3.2. **Bootstrap to assess variability in eigenvalue and subspace estimates.**

1. Compute and store the gradient samples $\nabla_{\mathbf{x}} f_1, \ldots, \nabla_{\mathbf{x}} f_M$ and the estimates $\hat{\Lambda}$ and $\hat{\mathbf{W}}$ from Algorithm 3.1.

2. Choose M_{boot} to be the number of bootstrap replicates—typically between 100 and 10000. For i from 1 to M_{boot}, do the following.

   (a) Draw a random integer j_k between 1 and M for $k = 1, \ldots, M$.

   (b) Compute the bootstrap replicate $\hat{C}_i^*$ as

   $$\hat{C}_i^* = \frac{1}{M} \sum_{k=1}^{M} (\nabla_{\mathbf{x}} f_{j_k})(\nabla_{\mathbf{x}} f_{j_k})^T. \qquad (3.72)$$

   (c) Compute the eigendecomposition $\hat{C}_i^* = \hat{\mathbf{W}}_i^* \hat{\Lambda}_i^* (\hat{\mathbf{W}}_i^*)^T$.

   (d) For a particular choice of the active subspace dimension n, compute

   $$e_i^* = \text{dist}(\text{ran}(\hat{\mathbf{W}}_1), \text{ran}(\hat{\mathbf{W}}_{i,1}^*)), \qquad (3.73)$$

   where $\hat{\mathbf{W}}_{i,1}^*$ contains the first n columns of $\hat{\mathbf{W}}_i^*$.

The bootstrap procedure computes M_{boot} eigenvalue decompositions. The ensemble of bootstrapped eigenvalues $\hat{\Lambda}_i^*$ can help assess the variability in the estimates. Similarly, the quantities e_i^* can be used to assess the variability in the subspace estimates.

As seen in Corollary 3.10, the error in the estimated subspace depends inversely on the gap between the eigenvalues scaled by the largest eigenvalue. The key to accurately approximating the subspace is to look for gaps in the eigenvalues; this is consistent with standard perturbation theory for eigenvector computations [116]. For example, if there is a larger gap between the third and fourth eigenvalues than between the second and third, then estimates of the three-dimensional subspace are more accurate than estimates of the two-dimensional subspace. This contrasts heuristics for deciding the dimension of the subspace in (i) model reduction based on the proper orthogonal decomposition [8] and (ii) dimension reduction based on principal component analysis [71]. In these cases, one chooses the dimension of the subspace by a threshold on the magnitude of the eigenvalues, e.g., so that the sum of retained eigenvalues exceeds some proportion of the sum of all eigenvalues. To accurately approximate the active subspace, the most important quantity is the spectral gap, which indicates a separation between subspaces.

To tease out the spectral gap, plot the estimated eigenvalues and their respective upper and lower bootstrap bounds; a gap between subsequent intervals provides strong evidence of a spectral gap when there is negligible error in the gradients. In sections 3.4.1 and 3.4.2, we show several examples of such plots (see Figures 3.2, 3.3, 3.5, and 3.6).

The bootstrap also helps us study the variability in the subspace estimate. In particular, for each replicate $\hat{\boldsymbol{W}}_i^*$ of the eigenvectors, we compute $\mathrm{dist}(\mathrm{ran}(\hat{\boldsymbol{W}}_1), \mathrm{ran}(\hat{\boldsymbol{W}}_{i,1}^*))$, where $\hat{\boldsymbol{W}}_{i,1}^*$ contains the first n columns of $\hat{\boldsymbol{W}}_i^*$. The bootstrap intervals on this distance help us to assess the subspace's stability. Recall that the distance between subspaces is bounded above by 1, so a bootstrap interval whose upper value is 1 indicates a poorly approximated active subspace. Figures 3.2, 3.4, 3.5, and 3.6 show examples of plotting this metric for the stability of the subspace; the first two figures also compare the measure of stability to the true error in the active subspace.

We summarize the practical approach to approximating the active subspace with bootstrap intervals. What follows is a modification of Algorithm 3.1 including our suggestions for parameter values. This procedure assumes the user has decided on the number k of eigenvalues to examine.

ALGORITHM **3.3. Monte Carlo estimation of the active subspace with bootstrap intervals.**

1. Choose $M = \alpha k \log(m)$, where α is a multiplier between 2 and 10.

2. Draw M samples $\{\mathbf{x}_j\}$ independently according to ρ, and for each $\mathbf{x}_j$ compute $\nabla_{\mathbf{x}} f_j = \nabla_{\mathbf{x}} f(\mathbf{x}_j)$.

3. Compute

$$\hat{C} = \frac{1}{M} \sum_{j=1}^{M} (\nabla_{\mathbf{x}} f_j)(\nabla_{\mathbf{x}} f_j)^T = \hat{\boldsymbol{W}} \hat{\Lambda} \hat{\boldsymbol{W}}^T. \tag{3.74}$$

4. Use the bootstrap (Algorithm 3.2) to compute replicates

$$\hat{C}_i^* = \hat{\boldsymbol{W}}_i^* \hat{\Lambda}_i^* (\hat{\boldsymbol{W}}_i^*)^T. \tag{3.75}$$

5. Plot the eigenvalues' bootstrap intervals and look for large gaps.

6. Choose the dimension n of the active subspace corresponding to the largest eigenvalue gap.

7. Use the bootstrap replicates of the eigenvectors to compute $\mathrm{dist}(\mathrm{ran}(\hat{\boldsymbol{W}}_1), \mathrm{ran}(\hat{\boldsymbol{W}}_{i,1}^*))$, and compute bootstrap intervals of this quantity.

A few comments are in order. First, if there is no perceivable gap in the eigenvalues, then an active subspace may not be present in the first $k-1$ dimensions. Repeating the process with larger k and more gradient samples may or may not reveal a gap in later eigenvalues. Second, if the bootstrap intervals on the subspace distance are not sufficiently less than 1, then the estimated subspace might be a poor approximation.

Third, we choose the bootstrap to examine the variability because we assume that sampling more gradients is too expensive. If this is not the case, i.e., if one can cheaply evaluate many more gradient samples, then one could use simple Monte Carlo in place of the bootstrap. We also assume that the dimension m of C is small enough that eigendecompositions of $\hat{C}$ and its bootstrap replicates are cheaper than sampling more gradients.

Lastly, we note that the elements of C are multivariate integrals. If m is small enough (2 or 3) and evaluating $\nabla_{\mathbf{x}} f$ is cheap enough, then more accurate numerical quadrature rules perform better than the random sampling. However, practical error estimates are more difficult to compute.

3.4.1 ▪ Example: A quadratic model

We apply Algorithm 3.3 to a quadratic function in 10 variables. For this case, we know the true eigenvalues and eigenvectors before we start estimating. Define f as

$$f(\mathbf{x}) = \frac{1}{2}\mathbf{x}^T A \mathbf{x}, \quad \mathbf{x} \in [-1,1]^{10}, \tag{3.76}$$

where A is symmetric and positive definite. We take $\rho = 2^{-10}$ on the hypercube $[-1,1]^{10}$ and zero elsewhere. The gradient is $\nabla_{\mathbf{x}} f(\mathbf{x}) = A\mathbf{x}$, so

$$C = A\left(\int \mathbf{x}\mathbf{x}^T \rho \, d\mathbf{x}\right) A^T = \frac{1}{3}A^2. \tag{3.77}$$

The eigenvalues of C are the squared eigenvalues of A, and the eigenvectors of C are the eigenvectors of A.

We study three different A's constructed from three choices for the eigenvalues: (i) exponential decay with a constant rate, (ii) exponential decay with a larger gap between the first and second eigenvalues, and (iii) exponential decay with a larger gap between the third and fourth eigenvalues. The three cases of A's eigenvalues are shown in Figures 3.2(a)–(c). Each A has the same eigenvectors, which we generate as an orthogonal basis from a random 10×10 matrix. In Figures 3.2–3.4, we refer to the three A's as cases 1, 2, and 3.

To estimate the eigenvalues, we choose M as in (3.71) with the multiplier $\alpha = 2$ and $k = 6$ eigenvalues of interest, which yields $M = 28$ evaluations of the gradient. Figures 3.2(d)–(f) show the bootstrap intervals for the first six eigenvalues along with the true eigenvalues of C. The tight intervals suggest confidence in the estimates. The gaps are apparent in the last two cases. Figures 3.2(g)–(i) show bootstrap intervals for the distance between the true k-dimensional active subspace and the subspace estimated with the M samples; the true distance is shown by the circles. Notice that subspaces corresponding to the larger eigenvalue gap are much better approximated than the others. For example, the three-dimensional subspace is better approximated than the one- and two-dimensional subspaces for the third case.

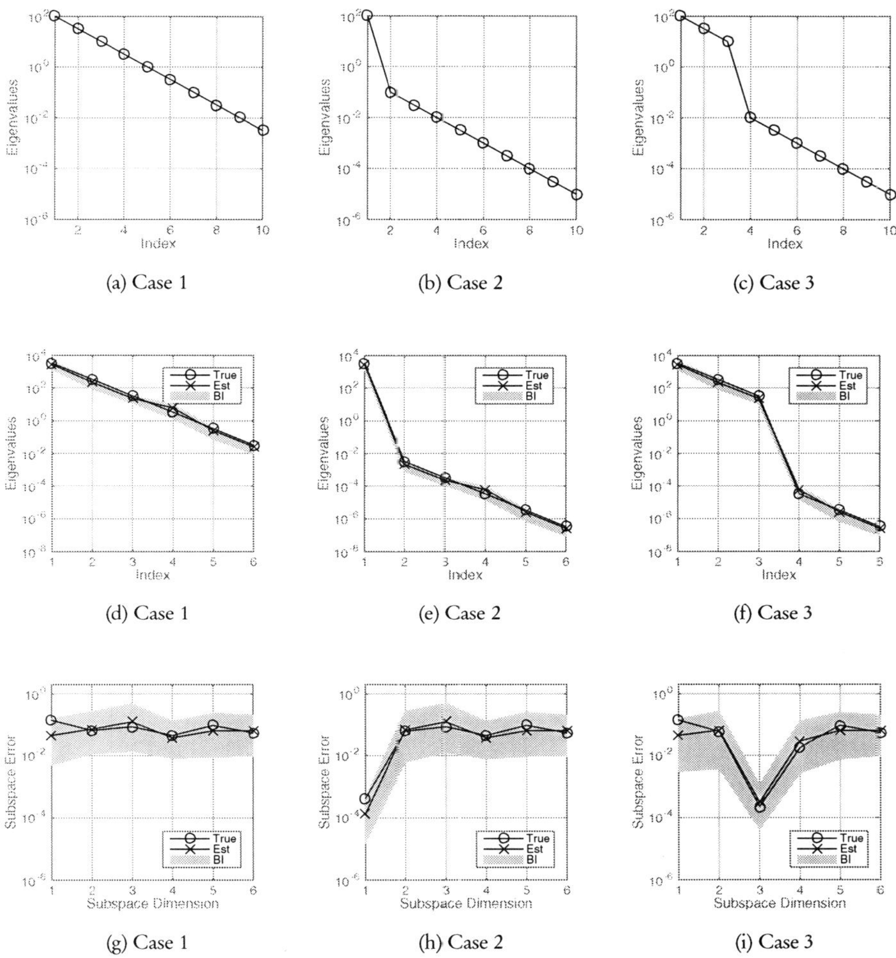

Figure 3.2. (a)–(c) *show the eigenvalues of the three choices for* **A**. (d)–(f) *show the true and estimated eigenvalues along with the bootstrap intervals; eigenvalues are well approximated for all three cases.* (g)–(i) *show the distance between the estimated subspace and the true subspace. In practice we do not have the true subspace, but we can estimate the distance with a bootstrap procedure as described in section* 3.4.

Next we repeat the study using first-order finite difference gradients with $h = 10^{-1}$, 10^{-3}, and 10^{-5}. With this gradient approximation, our error model (3.54) becomes $\gamma_h = \mathcal{O}(h)$. The first stepsize is large; we chose this large value to study the interplay between inaccurate gradients and the finite sample approximations of the eigenpairs. Figure 3.3 shows the true eigenvalues, their estimates, and the bootstrap intervals for all three cases and all three values of h; the horizontal lines show the value of h. Eigenvalues that are smaller than h are estimated less accurately than those larger than h, which is consistent with Theorem 3.12. Also the gaps are much less noticeable in the estimates when h is not small enough to resolve the smaller eigenvalue in the gapped pair. In particular, if the smaller eigenvalue in a gap is smaller than the finite difference stepsize, then a phantom eigenvalue appears; see Figures 3.3(b),(c) for examples of this phenomenon.

Figure 3.4 shows the distance between the true active subspace and the finite sample estimate with approximate gradients (circles). We use the bootstrap to estimate the error in the subspace as in section 3.4. There is a strong bias in the estimates of the subspace error when the phantom eigenvalues appear. For instance, in Figure 3.4, the estimates of the error for subspaces of dimension 4 through 6 are biased for $h = 10^{-3}$ and significantly biased for $h = 10^{-1}$. Compare this to the error in the last three eigenvalues in Figure 3.3.

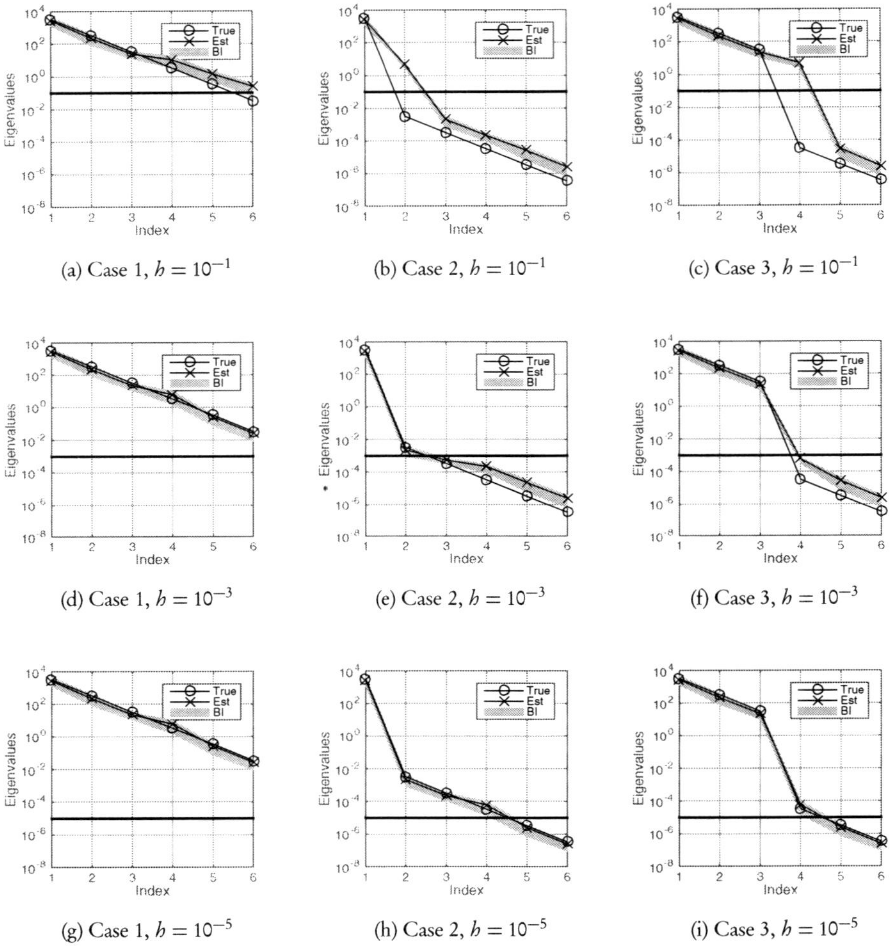

(a) Case 1, $h = 10^{-1}$ (b) Case 2, $h = 10^{-1}$ (c) Case 3, $h = 10^{-1}$

(d) Case 1, $h = 10^{-3}$ (e) Case 2, $h = 10^{-3}$ (f) Case 3, $h = 10^{-3}$

(g) Case 1, $h = 10^{-5}$ (h) Case 2, $h = 10^{-5}$ (i) Case 3, $h = 10^{-5}$

Figure 3.3. *Eigenvalues, estimates, and bootstrap intervals using finite difference gradients with $h = 10^{-1}$ ((a)–(c)), $h = 10^{-3}$ ((d)–(f)), and $h = 10^{-5}$ ((g)–(i)). The horizontal black lines indicate the value of h in each plot. In general, estimates of eigenvalues smaller than h are less accurate than those larger than h.*

3.4.2 ▪ Revisiting the parameterized PDE

In Chapter 1 we searched for an active subspace in a function derived from an idealized single-phase subsurface flow PDE model with uncertain permeability. We now provide details of this model, and we show results that include the bootstrap intervals for two

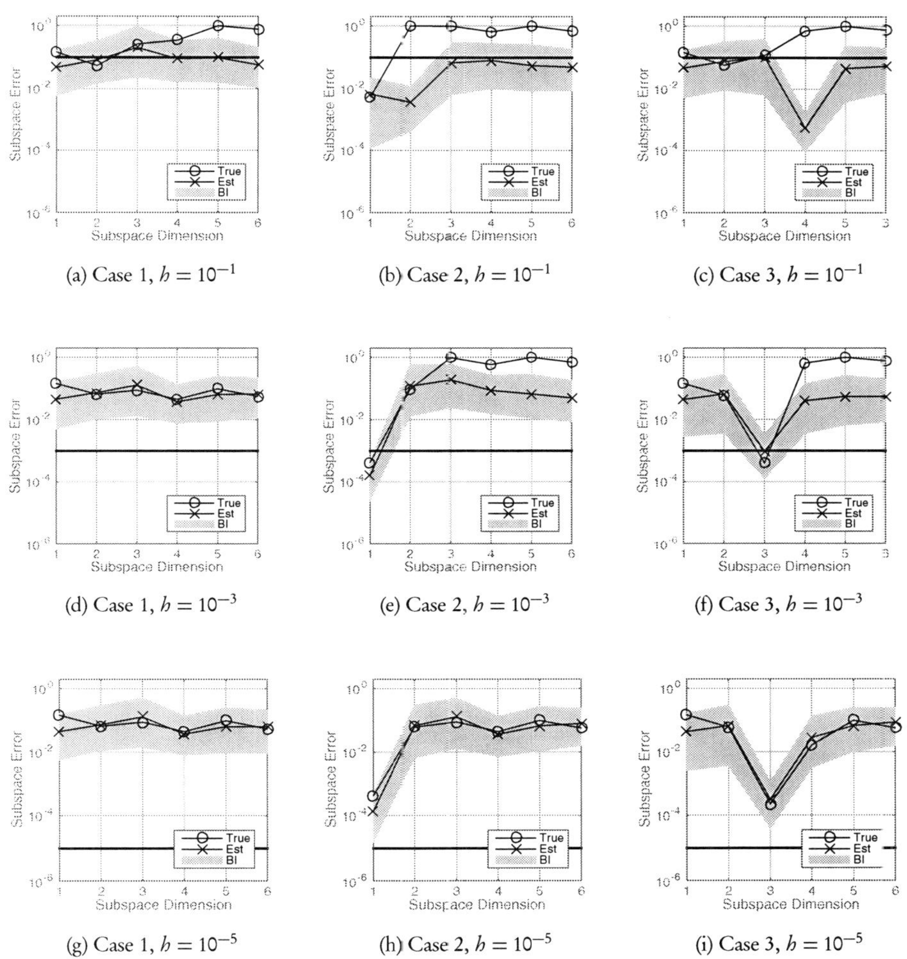

(a) Case 1, $h = 10^{-1}$ (b) Case 2, $h = 10^{-1}$ (c) Case 3, $h = 10^{-1}$

(d) Case 1, $h = 10^{-3}$ (e) Case 2, $h = 10^{-3}$ (f) Case 3, $h = 10^{-3}$

(g) Case 1, $h = 10^{-5}$ (h) Case 2, $h = 10^{-5}$ (i) Case 3, $h = 10^{-5}$

Figure 3.4. *The distance between the true active subspace and its finite sample approximation with finite difference gradients for $h = 10^{-1}$ ((a)–(c)), $h = 10^{-3}$ ((d)–(f)), and $h = 10^{-5}$ ((g)–(i)). The subspaces are very poorly approximated when h is not small enough to resolve the eigenvalues corresponding to the subspaces; compare to Figure 3.3.*

different correlation lengths in the random field modeling the coefficients; the results in Chapter 1 used the shorter correlation length. Consider the following Poisson equation with parameterized, spatially varying coefficients. Let $u = u(\mathbf{s}, \mathbf{x})$ satisfy

$$-\nabla_{\mathbf{s}} \cdot (a\,\nabla_{\mathbf{s}} u) = 1, \qquad \mathbf{s} \in [0,1]^2, \tag{3.78}$$

where $\mathbf{s}$ is the vector of two spatial coordinates. We set homogeneous Dirichlet boundary conditions on the left, top, and bottom of the spatial domain; denote this boundary by Γ_1. The right side of the spatial domain denoted by Γ_2 has a homogeneous Neumann boundary condition. The log of the coefficients $a = a(\mathbf{s}, \mathbf{x})$ is given by a truncated Karhunen–

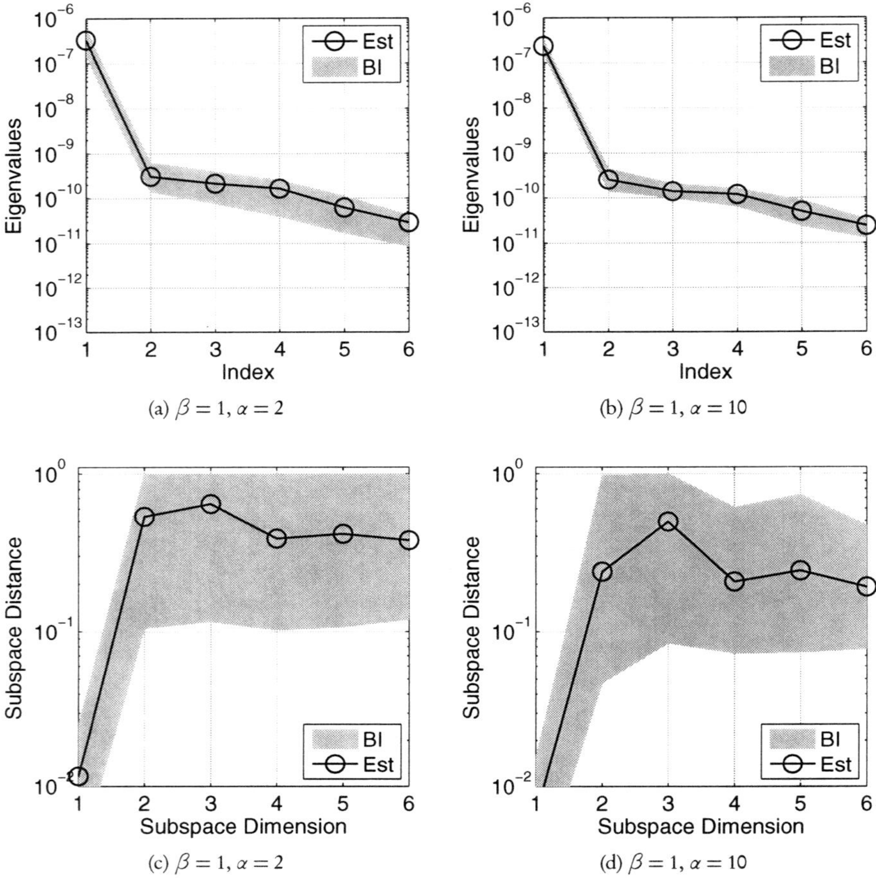

Figure 3.5. (a) *and* (b) *show estimates of the eigenvalues of* C *along with the bootstrap intervals for the quantity of interest* (3.81) *from the parameterized PDE model with the long correlation length* $\beta = 1$ *from* (3.80). (c) *and* (d) *show the estimates and bootstrap intervals on the distance between the estimated active subspace and the true active subspace.* (a) *and* (c) *are computed with the multiplier* $\alpha = 2$ *when choosing* M; (b) *and* (d) *use* $\alpha = 10$. *The gap between the first and second eigenvalues is significant as judged by the gap between the bootstrap intervals.*

Loève-type expansion [85],

$$\log(a(\mathbf{s}, \mathbf{x})) = \sum_{i=1}^{m} x_i \, \gamma_i \, \phi_i(\mathbf{s}), \tag{3.79}$$

where the x_i's are independent standard normal random variables, and the pairs $\{\gamma_i, \phi_i(\mathbf{s})\}$ are the eigenpairs of the correlation operator

$$\mathscr{C}(\mathbf{s}, \mathbf{t}) = \exp\left(-\beta^{-1} \|\mathbf{s} - \mathbf{t}\|_1\right). \tag{3.80}$$

We study the quality of the active subspace approximation for two correlation lengths, $\beta = 1$ and $\beta = 0.01$. These correspond to *long* and *short* correlation lengths, respectively, for the random field defining the log of the coefficients. We truncate the Karhunen–Loève series at $m = 100$ terms, which implies that the parameter space is $\mathscr{X} = \mathbb{R}^{100}$ with ρ a

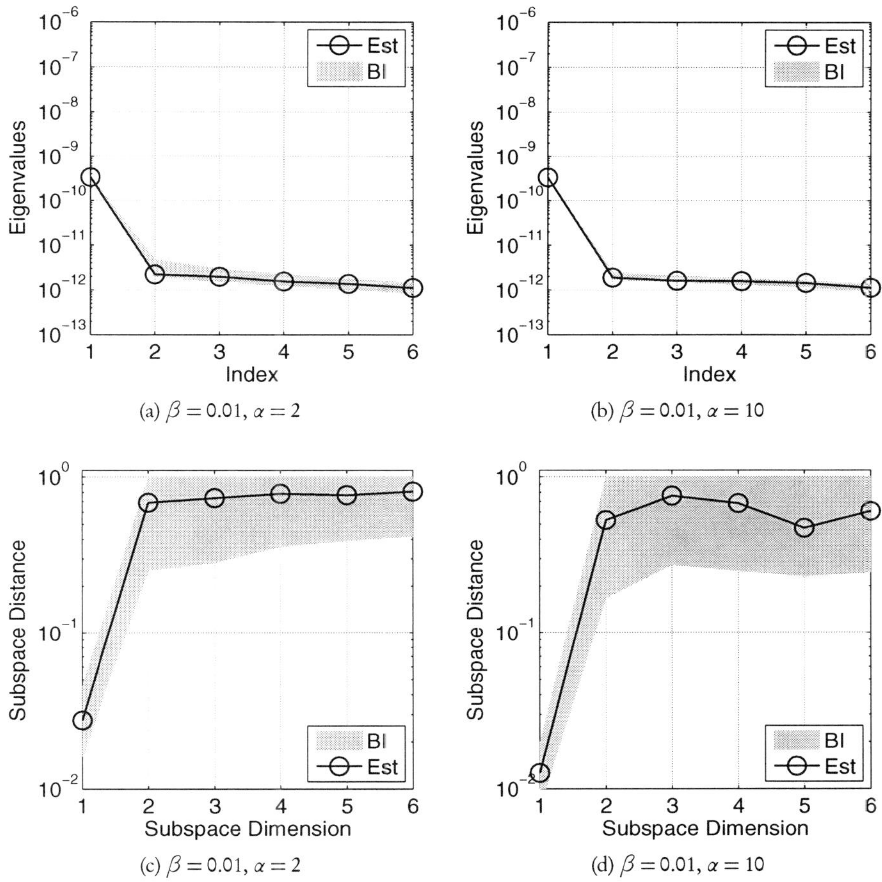

Figure 3.6. *These plots mirror those of Figure 3.5, except they use the short correlation length $\beta = 0.01$. The gap between the first and second eigenvalues is present in this case, too.*

standard Gaussian density function. Define the quantity of interest f to be the spatial average of the PDE solution over the boundary with the Neumann boundary condition,

$$f(\mathbf{x}) = \frac{1}{|\Gamma_2|} \int_{\Gamma_2} u(\mathbf{s}, \mathbf{x}) \, d\mathbf{s}. \tag{3.81}$$

Given a value for the input parameters $\mathbf{x}$, we discretize the PDE with a standard linear finite element method using the MATLAB PDE Toolbox. The discretized domain has 34320 triangles and 17361 nodes; the eigenfunctions ϕ_i from (3.79) are approximated on this mesh. We compute the gradient of the quantity of interest (3.81) using a discrete adjoint formulation. Further details appear in our previous work [28].

Figures 3.5(a),(b) show the estimates of the eigenvalues of C along with the bootstrap intervals for $\beta = 1$ in (3.80). There is a gap between the first and second eigenvalues, and the bootstrap intervals provide evidence that this gap is real—assuming the gradients are sufficiently accurate. We exploit this gap in [28] to construct an accurate univariate response surface of the active variable. Figures 3.5(c),(d) show bootstrap intervals on the subspace error, which identify variability in the estimated subspace. Figures 3.5(a),(c) show results with $M = 56$ gradient samples, which corresponds to a multiplier $\alpha = 2$. (b) and (d) use $M = 277$ samples, which corresponds to $\alpha = 10$. Notice that the bootstrap intervals for the eigenvalues decrease with more gradient samples. The error in the estimated subspace also decreases with more samples, although not substantially. Figure 3.6 shows the identical study with the short correlation length $\beta = 0.01$ from (3.80).

Chapter 4

Exploit the Active Subspace

If we have discovered an active subspace for a simulation's $f(\mathbf{x})$, then we want to exploit it to perform the parameter studies—optimization, averaging, inversion, or response surface construction—more efficiently. The general strategy is to focus a method's efforts on the active variables and pay less attention to the inactive variables. But the details differ for each parameter study. In this chapter we set up a framework for thinking about the dimension reduction, and we address each parameter study individually. The techniques for working with response surfaces are the most mature; these can be found in our prior work [28]. The remaining parameter studies have several open questions. This chapter is full of emerging ideas, and we look forward to their continued development.

4.1 ▪ Dimension reduction and mappings

We want to perform the parameter studies from section 2.2 using $f(\mathbf{x})$ and its gradient $\nabla_{\mathbf{x}} f(\mathbf{x})$ as a black box. In other words, we have a computer code that returns f and $\nabla_{\mathbf{x}} f$ given $\mathbf{x}$, and that code is usually very expensive; we want to use as few calls to the black box as possible. Unfortunately, the parameter study's methods may need an exorbitant number of calls to produce accurate results when m is large—like something that scales exponentially with m. The active subspace provides a structure we can exploit in the parameter studies, namely, by focusing the methods' efforts on the active variables.

We want to construct a function $z = g(\mathbf{y})$ that depends only on the $n < m$ active variables (3.9) and study its parametric dependence as a proxy for $f(\mathbf{x})$. And we would like to treat g like a black box, just like f. If we can find such a g, then we can perform the parameter studies in $n < m$ variables—potentially using exponentially fewer black box calls. Finally, we want to translate the parameter studies' results from $g(\mathbf{y})$ to $f(\mathbf{x})$. But constructing g might be tricky: how do we build a function that depends on linear combinations of the simulation's inputs $\mathbf{x}$? Such a function may not make physical sense. A black box call to g must connect to $f(\mathbf{x})$, since that is the interface we have available.

Subset selection approaches to dimension reduction use measures of global sensitivity to ignore unimportant variables, e.g., variables that do not contribute to the variance of f. They may ignore unimportant variables by fixing them at nominal values and letting the important variables vary. We can try a similar approach. Recall that

$$f(\mathbf{x}) = f(W_1 \mathbf{y} + W_2 \mathbf{z}) \tag{4.1}$$

and that small perturbations in $\mathbf{z}$ change f relatively little, on average. We could fix $\mathbf{z} = 0$, which corresponds to its nominal value, and construct g as

$$g(\mathbf{y}) = f(\mathbf{W}_1\mathbf{y}). \tag{4.2}$$

This construction connects the black box g to the black box f. Given a value for $\mathbf{y}$, first set $\mathbf{x} = \mathbf{W}_1\mathbf{y}$, and then evaluate $f(\mathbf{x})$. If we wanted to use this g as a proxy for f, then we could try

$$f(\mathbf{x}) \approx g(\mathbf{W}_1^T\mathbf{x}) = f(\mathbf{W}_1\mathbf{W}_1^T\mathbf{x}). \tag{4.3}$$

In other words, this proxy evaluates f at the projection of $\mathbf{x}$ onto the subspace spanned by the columns of $\mathbf{W}_1$—and this subspace is precisely the active subspace. In fact, if the eigenvalues $\lambda_{n+1},\ldots,\lambda_m$ from (3.5) are exactly zero, and if f's domain $\mathcal{X}$ is $\mathbb{R}^m$, then this is an excellent choice for g. All parameter studies can be confined to the inputs in the active subspace, which has dimension $n < m$. But let's see what happens when either of these conditions is not met.

Suppose that λ_{n+1} is not exactly zero. Fix $\mathbf{y}$. Lemma 3.1 suggests that the squared finite difference approximation of the derivative along $\mathbf{w}_{n+1}$,

$$\frac{1}{\delta^2}\left(f(\mathbf{W}_1\mathbf{y} + \delta\mathbf{w}_{n+1}) - f(\mathbf{W}_1\mathbf{y})\right)^2, \tag{4.4}$$

is, on average, close to λ_{n+1} for sufficiently small δ. But for larger δ, the eigenvalue λ_{n+1} does not give much information. If the difference between $f(\mathbf{W}_1\mathbf{y})$ and $f(\mathbf{W}_1\mathbf{y} + \delta\mathbf{w}_{n+1})$ is $\mathcal{O}(\delta)$, then large δ implies a large difference between the function values. In other words, the distance between $\mathbf{x}$ and its active subspace projection $\mathbf{W}_1\mathbf{W}_1^T\mathbf{x}$ may affect the approximation accuracy of $g(\mathbf{y})$ constructed as (4.2). We want something more robust.

Next, consider the function from Figure 1.1, $f(\mathbf{x}) = \exp(\mathbf{a}^T\mathbf{x})$ with $\mathbf{a} = [0.7, 0.3]^T$ defined on the box $\mathbf{x} \in [-1,1]^2$. This f has the structure discussed in section 3.2.1. The active subspace is defined by the 2×1 matrix $\mathbf{W}_1 = \mathbf{a}/\|\mathbf{a}\|$. The projection of the point $[0.9, 0.9]^T$ onto the active subspace is

$$(\mathbf{W}_1\mathbf{W}_1^T)\begin{bmatrix}0.9\\0.9\end{bmatrix} = \begin{bmatrix}1.09\\0.47\end{bmatrix}. \tag{4.5}$$

The first component of this projection is greater than 1, so the projection of $\mathbf{x}$ onto the active subspace is outside the domain of f. This happens with several points in the domain because of the bounds. As a general rule, we want to construct a g that avoids evaluating f outside its domain.

These two issues demonstrate an important point. The relationship between the range of the columns of $\mathbf{W}_1$ and f's inputs is subtle. The active subspace is a set of directions. Perturbations in the inputs along these directions change f more, on average, than perturbations orthogonal to these directions. Except for special cases—namely, $\mathcal{X} = \mathbb{R}^m$ and $\lambda_{n+1} = \cdots = \lambda_m = 0$—the active subspace does not restrict our attention to the intersection of $\text{ran}(\mathbf{W}_1)$ and $\mathcal{X}$. To construct g, we need to respect this subtlety.

We first carefully define the domain of g. Let $\mathcal{Y} \subseteq \mathbb{R}^n$ be the set

$$\mathcal{Y} = \left\{\mathbf{y} \in \mathbb{R}^n,\, \mathbf{y} = \mathbf{W}_1^T\mathbf{x},\, \mathbf{x} \in \mathcal{X}\right\}. \tag{4.6}$$

This set is the domain of $g(\mathbf{y})$. Any point $\mathbf{y} \in \mathcal{Y}$ is guaranteed to have at least one $\mathbf{x} \in \mathcal{X}$ such that $\mathbf{y} = \mathbf{W}_1^T\mathbf{x}$. Then we have at least one choice for $g(\mathbf{y})$: given $\mathbf{y} \in \mathcal{Y}$, find $\mathbf{x} \in \mathcal{X}$ such that $\mathbf{y} = \mathbf{W}_1^T\mathbf{x}$, and set $g(\mathbf{y}) = f(\mathbf{x})$. The trouble is that there are likely infinitely many $\mathbf{x}$ such that $\mathbf{y} = \mathbf{W}_1^T\mathbf{x}$, i.e., the map from $\mathbf{y}$ to $\mathbf{x}$ is ill-posed. And each $\mathbf{x}$ maps to a particular $f(\mathbf{x})$. How does one choose to set $g(\mathbf{y})$ among infinitely many $f(\mathbf{x})$'s? To overcome the ill-posedness, we must choose a *regularization*.

The choice may not matter. If $\lambda_{n+1} = \cdots = \lambda_m = 0$, then $f(\mathbf{x})$ is the same for *any* $\mathbf{x}$ such that $\mathbf{y} = \mathbf{W}_1^T \mathbf{x}$. In this case, the map from $\mathbf{y}$ to $g(\mathbf{y})$ can be uniquely defined with an arbitrary regularization; choose one that is computationally convenient. But this is a special case.

In general, when the eigenvalues associated with the inactive subspace are small but not zero, the choice of regularization depends on the parameter study. We motivate and develop the following regularization choices in subsequent sections.

1. For response surfaces, integration, and statistical inversion, we define $g(\mathbf{y})$ by *averaging* over all $f(\mathbf{x})$ such that $\mathbf{y} = \mathbf{W}_1^T \mathbf{x}$. See sections 4.2, 4.3, and 4.5, respectively.

2. For optimization, we define $g(\mathbf{y})$ by *optimizing* over all $f(\mathbf{x})$ such that $\mathbf{y} = \mathbf{W}_1^T \mathbf{x}$. See section 4.4.

The first choice sets $g(\mathbf{y})$ to be the conditional expectation of f given y. This regularization lets us relate the error in g to the eigenvalues $\lambda_{n+1}, \ldots, \lambda_m$. If perturbations in the inactive variables change f relatively little, then we can average over the inactive variables with relatively little effort. The second choice distinguishes the active from the inactive variables for optimization. A unit of effort to optimize over $\mathbf{y}$ should on average produce greater change than the same effort to optimize over $\mathbf{z}$. Regularization for g that limits efforts to optimize over $\mathbf{z}$—say, through cheap models or a fixed number of iterations— and emphasizes efforts to optimize over $\mathbf{y}$ may yield an efficient strategy for optimizing f.

To define things like integration and averaging, we need to translate the probability density $\rho = \rho(\mathbf{x})$ into comparable quantities for the active and inactive variables. The joint density function of the active variables $\mathbf{y}$ and the inactive variables $\mathbf{z}$ is

$$\pi(\mathbf{y}, \mathbf{z}) = \rho(\mathbf{W}_1 \mathbf{y} + \mathbf{W}_2 \mathbf{z}). \tag{4.7}$$

The marginal densities are

$$\pi_Y(\mathbf{y}) = \int \rho(\mathbf{W}_1 \mathbf{y} + \mathbf{W}_2 \mathbf{z}) \, d\mathbf{z}, \qquad \pi_Z(\mathbf{z}) = \int \rho(\mathbf{W}_1 \mathbf{y} + \mathbf{W}_2 \mathbf{z}) \, d\mathbf{y}, \tag{4.8}$$

and the conditional densities are

$$\pi_{Y|Z}(\mathbf{y}|\mathbf{z}) = \frac{\pi(\mathbf{y}, \mathbf{z})}{\pi_Z(\mathbf{z})}, \qquad \pi_{Z|Y}(\mathbf{z}|\mathbf{y}) = \frac{\pi(\mathbf{y}, \mathbf{z})}{\pi_Y(\mathbf{y})}. \tag{4.9}$$

These densities play a role in the response surface, integration, and inversion parameter studies.

We are especially concerned with two particular cases frequently found in practice. We note when results are specific to these cases. The first case is when the domain $\mathcal{X} = \mathbb{R}^m$ is unbounded and $\rho(\mathbf{x})$ is a standard Gaussian density function. This case avoids several issues caused by bounded domains. The domain $\mathcal{Y}$ of g is also unbounded and equal to $\mathbb{R}^n$. Also, since the columns of $\mathbf{W}$ are orthogonal, the marginal and conditional densities of $\mathbf{y}$ are standard Gaussian densities on $\mathbb{R}^n$, and the marginal and conditional densities of $\mathbf{z}$ are standard Gaussian densities on $\mathbb{R}^{m-n}$.

The second case is when the domain $\mathcal{X} = [-1, 1]^m$ is a hypercube and the probability density is uniform:

$$\rho(\mathbf{x}) = \begin{cases} 2^{-m}, & \mathbf{x} \in [-1, 1]^m, \\ 0 & \text{otherwise.} \end{cases} \tag{4.10}$$

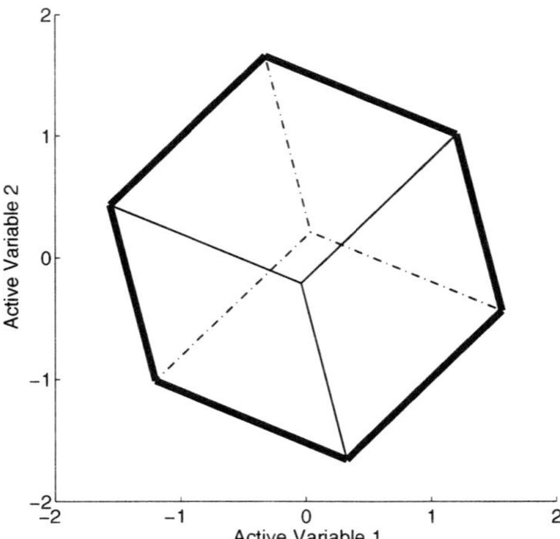

Figure 4.1. *A three-dimensional cube* $[-1, 1]^3$ *rotated and photographed. The dotted lines show the cube's edges in the background. The thick lines show the boundary of the two-dimensional zonotope. The axes are labeled for the first and second active variables.*

This case is considerably more complex. The domain $\mathcal{Y}$ is the image of the hypercube under an orthogonal projection. This object is a convex polytope in $\mathbb{R}^n$ known as a *zonotope* [53]. Consider randomly rotating a three-dimensional cube and taking a photograph as in Figure 4.1. The outline of the cube in the photograph has, in general, six vertices and six sides. The two remaining vertices from the cube are inside the convex hull of the six exterior vertices. Zonotopes generalize this idea to higher dimensions. The marginal and conditional densities are not necessarily uniform, and they can get very complicated in high dimensions. To see that these densities are not generally uniform, let $m = 2$, and assume the eigenvectors from (3.5) are given by

$$W = \begin{bmatrix} a & -b \\ b & a \end{bmatrix}, \quad a, b > 0, \quad \sqrt{a^2 + b^2} = 1. \tag{4.11}$$

Take $y = ax_1 + bx_2$ and $z = -bx_1 + ax_2$. The range of y is the interval $[-(a + b), a + b]$, and the marginal density of y is

$$\pi_Y(y) = \frac{z_u(y) - z_\ell(y)}{4}, \tag{4.12}$$

where

$$z_\ell(y) = \max\left(\frac{1 - ay}{-b}, \frac{-1 - by}{a} \right),$$

$$z_u(y) = \min\left(\frac{-1 - ay}{-b}, \frac{1 - by}{a} \right). \tag{4.13}$$

When $m > 2$ and $n > 1$, these constructions become much more complicated.

4.2 ▪ Response surfaces

We use *response surface* as a catch-all term for the following generic procedure. First, choose a set of points $\{x_i\}$ with each $x_i \in \mathcal{X}$, and compute $q_i = f(x_i)$. Then use the pairs $\{\mathbf{x}_i, q_i\}$ to approximate $f(\mathbf{x})$ for some $\mathbf{x} \notin \{\mathbf{x}_i\}$. If the approximation equals the true function at the $\mathbf{x}_i$'s, then the response surface interpolates the q_i's. Common interpolation procedures in several variables use radial basis functions [124] or polynomials [56]. Response surfaces that do not interpolate are regressions, and they assume noise or error in the q_i's [65]. These approximation procedures suffer from the curse of dimensionality, so they may benefit tremendously from proper dimension reduction.

The straightforward way to exploit the active subspace for response surfaces is to build a regression surface on the n active variables $\mathbf{y}$ instead of all m variables $\mathbf{x}$ as in the following algorithm.

ALGORITHM 4.1. **Constructing a regression surface on the active variables.**

1. Assume the pairs $\{\mathbf{x}_i, q_i\}$ are given.
2. For each $\mathbf{x}_i$, compute $\mathbf{y}_i = W_1^T \mathbf{x}_i$.
3. Fit a regression surface $g(\mathbf{y})$ with the pairs $\{\mathbf{y}_i, q_i\}$, where $q_i \approx g(\mathbf{y}_i)$.
4. Approximate
$$f(\mathbf{x}) \approx g(W_1^T \mathbf{x}). \tag{4.14}$$

The pairs $\{\mathbf{x}_i, q_i\}$ may be a leftover from the process of discovering the active subspace. The model in (4.14) is an n-index model [83] with a specific choice for the linear transformation, and the function g is the *link function*. The advantage of this construction is that more models are feasible—e.g., higher degree polynomials—for the n-variate function g than for directly approximating the m-variate function f. If one begins with 50 pairs $(\mathbf{x}_i, q_i)$ and $m = 10$, then fitting a quadratic surface with least-squares requires additional regularization. But if $n = 1$, then 50 pairs $(\mathbf{y}_i, q_i)$ are sufficient to fit a quadratic model with standard least-squares. One can judge the quality of the fitted regression surface with standard metrics such as the coefficient of determination.

There are two issues with this construction. First, if the $\mathbf{x}_i$'s were carefully chosen to satisfy some experimental design criteria, then the $\mathbf{y}_i$'s are unlikely to satisfy the same criteria. The $\mathbf{y}_i$'s may be bad points for response surface construction, even if the $\mathbf{x}_i$'s were good points. Second, fitting a regression surface assumes random error in the q_i's. In fact, the deviation of q_i from the low-dimensional model is not random; it is due to variation in the known, omitted inactive variables $\mathbf{z}$—assuming some of $\lambda_{n+1}, \ldots, \lambda_m$ are greater than zero. While the last point is somewhat philosophical, it may inhibit proper interpretation of the regression surface and its subsequent use.

In what follows, we develop a more careful response surface strategy that exploits the active subspace and is amenable to analysis. This development closely follows our paper [28], although the notation here is slightly different.

4.2.1 ▪ Conditional expectation and Monte Carlo approximation

For a fixed $\mathbf{y}$, the best guess one can make as to the value of f is its average over all values of $\mathbf{x}$ that map to $\mathbf{y}$; this is precisely the conditional expectation of f given $\mathbf{y}$. Define g by
$$g(\mathbf{y}) = \int f(W_1\mathbf{y} + W_2\mathbf{z})\, \pi_{Z|Y}(\mathbf{z})\, d\mathbf{z}. \tag{4.15}$$

The domain of this function is $\mathcal{Y}$ from (4.6). Since g is a conditional expectation, it is the best mean-squared approximation of f given $\mathbf{y}$ [126, Chapter 9].

We can use g to approximate f at a given $\mathbf{x}$ with the construction

$$f(\mathbf{x}) \approx g(\boldsymbol{W}_1^T \mathbf{x}). \tag{4.16}$$

To analyze the error in this construction, we need a particular Poincaré inequality that bounds the root-mean-square of a zero-mean function by the average norm of its gradient. The following two lemmas address the two primary cases of interest: the uniform case and the Gaussian case. The first lemma is a slight variation of the inequality from Bebendorf [4].

Lemma 4.1. *Assume that $\mathscr{X} \subset \mathbb{R}^m$ is convex and that the probability density function $\rho : \mathbb{R}^m \to \mathbb{R}_+$ is bounded, $\rho(\mathbf{x}) \leq R$, for all $\mathbf{x} \in \mathscr{X}$. Let h be a zero-mean, Lipschitz continuous, differentiable, real-valued function of $\mathbf{x}$. Then*

$$\int \big(h(\mathbf{x})\big)^2 \rho\, d\mathbf{x} \leq R\, C \int \big(\nabla_{\mathbf{x}} h(\mathbf{x})\big)^T \big(\nabla_{\mathbf{x}} h(\mathbf{x})\big) \rho\, d\mathbf{x}, \tag{4.17}$$

where C is the Poincaré constant for $\mathscr{X}$.

The novelty of Bebendorf's paper is that he derives an explicit estimate of the Poincaré constant in terms of the diameter of $\mathscr{X}$. If $\mathscr{X}$ is the hypercube $[-1,1]^m$ and ρ is the uniform density 2^{-m}, then C from Lemma 4.1 is bounded by $2\sqrt{m}/\pi$. The Poincaré inequality for the Gaussian case is due to Chen [26, Theorem 2.1], and we restate it here for completeness.

Lemma 4.2. *Assume that $\mathscr{X} = \mathbb{R}^m$, and $\rho(\mathbf{x})$ is a standard Gaussian density. Let h be a zero-mean, Lipschitz continuous, differentiable, real-valued function of $\mathbf{x}$. Then*

$$\int \big(h(\mathbf{x})\big)^2 \rho\, d\mathbf{x} \leq \int \big(\nabla_{\mathbf{x}} h(\mathbf{x})\big)^T \big(\nabla_{\mathbf{x}} h(\mathbf{x})\big) \rho\, d\mathbf{x}. \tag{4.18}$$

In other words, for the standard Gaussian density case, the constant is 1. Corollary 3.2 from Chen [26] extends this result to the case when $\mathbf{x}$ is a zero-mean Gaussian with a given covariance matrix. The next theorem estimates the error in (4.16) in terms of the eigenvalues of C from (3.2).

Theorem 4.3. *Under the conditions on $\mathscr{X}$ and ρ from either Lemma 4.1 or Lemma 4.2, the root-mean-squared error of the approximation defined in (4.16) satisfies*

$$\left(\int \big(f(\mathbf{x}) - g(\boldsymbol{W}_1^T \mathbf{x})\big)^2 \rho\, d\mathbf{x} \right)^{\frac{1}{2}} \leq C_1 (\lambda_{n+1} + \cdots + \lambda_m)^{\frac{1}{2}}, \tag{4.19}$$

where C_1 is a constant that depends on the domain $\mathscr{X}$ and the density function ρ.

Proof. Note that

$$\int \big(f(\boldsymbol{W}_1 \mathbf{y} + \boldsymbol{W}_2 \mathbf{z}) - g(\mathbf{y})\big) \pi_{Z|Y}(\mathbf{z})\, d\mathbf{z} = 0 \tag{4.20}$$

by the definition (4.16). Thus,

$$\int \left(f(\mathbf{x})-g(W_1^T\mathbf{x})\right)^2 \rho \, d\mathbf{x} \tag{4.21}$$

$$= \int \left(\int \left(f(W_1\mathbf{y}+W_2\mathbf{z})-g(\mathbf{y})\right)^2 \pi_{Z|Y}\, d\mathbf{z}\right) \pi_Y \, d\mathbf{y} \tag{4.22}$$

$$\leq C_1^2 \int \left(\int \left(\nabla_z f(W_1\mathbf{y}+W_2\mathbf{z})\right)^T\left(\nabla_z f(W_1\mathbf{y}+W_2\mathbf{z})\right)\pi_{Z|Y}\, d\mathbf{z}\right)\pi_Y\, d\mathbf{y} \tag{4.23}$$

$$= C_1^2 \int \left(\nabla_z f(\mathbf{x})\right)^T\left(\nabla_z f(\mathbf{x})\right)\rho\, d\mathbf{x} \tag{4.24}$$

$$= C_1^2 \left(\lambda_{n+1}+\cdots+\lambda_m\right). \tag{4.25}$$

Lines (4.22) and (4.24) are due to the tower property of conditional expectations [126]. Line (4.23) uses the Poincaré inequality from either Lemma 4.1 or Lemma 4.2, depending on $\mathcal{X}$ and ρ; the constant C_1^2 depends only on $\mathcal{X}$ and the density function ρ. Line (4.25) follows from Lemma 3.2. Taking the square root of both sides completes the proof. $\square$

The crux of the proof is the Poincaré inequality. However, it is well known that the Poincaré inequality is notoriously loose to account for the worst case; we expect that the error estimate from Theorem 4.3 can be improved with more assumptions on f.

The trouble with the approximation g from (4.16) is that each evaluation of g requires an integral with respect to the inactive variables $\mathbf{z}$. In other words, evaluating g requires high-dimensional integration. However, if the eigenvalues $\lambda_{n+1},\dots,\lambda_m$ are small, then small changes in $\mathbf{z}$ produce little change in f. Thus, the variance of f along $\mathbf{z}$ is small, and simple Monte Carlo will accurately approximate the conditional expectation g. We use simple Monte Carlo to approximate g and derive an error bound on such an approximation. The error bound validates the intuition that we need very few evaluations of f to approximate g if the eigenvalues are small.

Define the Monte Carlo estimate $\hat{g}=\hat{g}(\mathbf{y})$ by

$$g(\mathbf{y}) \approx \hat{g}(\mathbf{y}) = \frac{1}{N}\sum_{i=1}^{N} f(W_1\mathbf{y}+W_2\mathbf{z}_i), \tag{4.26}$$

where the $\mathbf{z}_i$ are drawn independently from the conditional density $\pi_{Z|Y}$. We approximate f as

$$f(\mathbf{x}) \approx \hat{g}(W_1^T\mathbf{x}). \tag{4.27}$$

Next we derive an error bound for this approximation.

Theorem 4.4. *The root-mean-squared error of $\hat{g}$ defined in (4.27) satisfies*

$$\left(\int (f(\mathbf{x})-\hat{g}(W_1^T\mathbf{x}))^2 \rho\, d\mathbf{x}\right)^{\frac{1}{2}} \leq C_1\left(1+N^{-\frac{1}{2}}\right)(\lambda_{n+1}+\cdots+\lambda_m)^{\frac{1}{2}}, \tag{4.28}$$

where C_1 is from Theorem 4.3.

Proof. First, define the conditional variance of f given $\mathbf{y}$ as

$$\sigma_\mathbf{y}^2 = \int (f(W_1\mathbf{y}+W_2\mathbf{z})-g(\mathbf{y}))^2 \pi_{Z|Y}\, d\mathbf{z}, \tag{4.29}$$

and note that the proof of Theorem 4.3 shows

$$\int \sigma_{\mathbf{y}}^2 \, \pi_Y \, d\mathbf{y} \leq C_1^2 \left(\lambda_{n+1} + \cdots + \lambda_m \right). \tag{4.30}$$

The Monte Carlo approximation $\hat{g}(\mathbf{y})$ is a random variable by virtue of the independent random samples $\mathbf{z}_i$, so we can take its expectation against $\pi_{Z|Y}$. The mean-squared error in $\hat{g}(\mathbf{y})$ satisfies [100]

$$\int (g(\mathbf{y}) - \hat{g}(\mathbf{y}))^2 \, \pi_{Z|Y} \, d\mathbf{z} = \frac{\sigma_{\mathbf{y}}^2}{N}, \tag{4.31}$$

so that

$$\begin{aligned}
\int (g(\boldsymbol{W}_1^T \mathbf{x}) - \hat{g}(\boldsymbol{W}_1^T \mathbf{x}))^2 \, \rho \, d\mathbf{x} \\
= \int \left(\int (g(\mathbf{y}) - \hat{g}(\mathbf{y}))^2 \pi_{Z|Y} \, d\mathbf{z} \right) \pi_Y \, d\mathbf{y} \\
= \frac{1}{N} \int \sigma_{\mathbf{y}}^2 \, \pi_Y \, d\mathbf{y} \\
\leq \frac{C_1^2}{N} \left(\lambda_{n+1} + \cdots + \lambda_m \right).
\end{aligned} \tag{4.32}$$

Finally, using Theorem 4.3,

$$\begin{aligned}
\left(\int (f(\mathbf{x}) - \hat{g}(\boldsymbol{W}_1^T \mathbf{x}))^2 \, \rho \, d\mathbf{x} \right)^{\frac{1}{2}} \\
\leq \left(\int (f(\mathbf{x}) - g(\boldsymbol{W}_1^T \mathbf{x}))^2 \, \rho \, d\mathbf{x} \right)^{\frac{1}{2}} + \left(\int (g(\boldsymbol{W}_1^T \mathbf{x}) - \hat{g}(\boldsymbol{W}_1^T \mathbf{x}))^2 \, \rho \, d\mathbf{x} \right)^{\frac{1}{2}} \\
\leq C_1 \left(1 + N^{-\frac{1}{2}} \right) \left(\lambda_{n+1} + \cdots + \lambda_m \right)^{\frac{1}{2}},
\end{aligned} \tag{4.33}$$

as required.   □

If $\lambda_{n+1} = \cdots = \lambda_m = 0$, then the Monte Carlo estimate $\hat{g}$ is exact for any $N \geq 1$. The average of a constant function can be computed by evaluating the function once. More generally, if $\lambda_{n+1}, \ldots, \lambda_m$ are sufficiently small, then the Monte Carlo estimate $\hat{g}$ with small N produces a very good approximation of f.

We now reach the point where $n < m$ can reduce the cost of approximating f. Up to this point, we have seen no advantage to using the conditional expectation g or its Monte Carlo approximation $\hat{g}$ to approximate f; each evaluation of $\hat{g}(\mathbf{y})$ requires at least one evaluation of $f(\mathbf{x})$. The real advantage of the active subspace is that one can construct response surfaces with only the active variables $\mathbf{y} \in \mathbb{R}^n$ instead of f's natural variables $\mathbf{x} \in \mathbb{R}^m$. We train a response surface on the domain $\mathscr{Y}$ from (4.6) using a few evaluations of $\hat{g} = \hat{g}(\mathbf{y})$.

We do not specify the form of the response surface; several are possible, and our previous work [28, section 4] discusses applications using kriging. However, there is one important consideration before choosing a response surface method willy-nilly: if the eigenvalues $\lambda_{n+1}, \ldots, \lambda_m$ are not exactly zero, then the Monte Carlo estimate $\hat{g}$ contains *noise* due to the finite number of samples; in other words, $\hat{g}$ is a random variable. This noise implies that $\hat{g}$ is not a smooth function of $\mathbf{y}$—even if $g(\mathbf{y})$ is. Therefore, we prefer smoothing, regression-based response surfaces over exact interpolation.

We construct a generic response surface for a function defined on $\mathscr{Y}$ from (4.6) as follows. Define the *design* on $\mathscr{Y}$ to be a set of points $\mathbf{y}_k \in \mathscr{Y}$ with $k = 1,\dots,P$; we treat the design as fixed and not random. The specific design depends on the form of the response surface. Define $\hat{g}_k = \hat{g}(\mathbf{y}_k)$. Then we approximate

$$\hat{g}(\mathbf{y}) \approx \mathscr{R}(\mathbf{y}; \hat{g}_1,\dots,\hat{g}_P), \tag{4.34}$$

where $\mathscr{R}$ is a response surface constructed with the training data $\hat{g}_1,\dots,\hat{g}_P$. To keep the notation concise, we write $\mathscr{R}(\mathbf{y})$, where the dependence on the training data is implied. We use this response surface to approximate f as

$$f(\mathbf{x}) \approx \mathscr{R}(\boldsymbol{W}_1^T \mathbf{x}). \tag{4.35}$$

To derive an error estimate for $\mathscr{R}$, we assume that the average error in the response surface is bounded for all $\mathbf{y} \in \mathscr{Y}$.

Assumption 1. *There exists a constant C_2 such that*

$$\left(\int (\hat{g}(\mathbf{y}) - \mathscr{R}(\mathbf{y}))^2\, \pi_{Z|Y}\, d\mathbf{z} \right)^{\frac{1}{2}} \leq C_2 \delta \quad \text{for all } \mathbf{y} \in \mathscr{Y}, \tag{4.36}$$

where $\delta = \delta(\mathscr{R}, P)$ depends on the response surface method $\mathscr{R}$ and the number P of training data, and C_2 depends on the domain $\mathscr{X}$ and the probability density function ρ.

In theory, Assumption 1 is reasonable as long as the response surface construction is well-posed. For example, a polynomial surface fit with least-squares should be properly regularized. In practice, δ might be estimated from the minimum residuals.

Theorem 4.5. *The root-mean-squared error in $\mathscr{R}$ defined in (4.35) satisfies*

$$\left(\int (f(\mathbf{x}) - \mathscr{R}(\boldsymbol{W}_1^T \mathbf{x}))^2\, \rho\, d\mathbf{x} \right)^{\frac{1}{2}} \leq C_1 \left(1 + N^{-\frac{1}{2}} \right) \left(\lambda_{n+1} + \cdots + \lambda_m \right)^{\frac{1}{2}} + C_2 \delta, \tag{4.37}$$

where C_1 is from Theorem 4.3, N is from Theorem 4.4, and C_2 and δ are from Assumption 1.

Proof. Note that

$$\left(\int (f(\mathbf{x}) - \mathscr{R}(\boldsymbol{W}_1^T \mathbf{x}))^2\, \rho\, d\mathbf{x} \right)^{\frac{1}{2}}$$
$$\leq \left(\int (f(\mathbf{x}) - \hat{g}(\boldsymbol{W}_1^T \mathbf{x}))^2\, \rho\, d\mathbf{x} \right)^{\frac{1}{2}} + \left(\int (\hat{g}(\boldsymbol{W}_1^T \mathbf{x}) - \mathscr{R}(\boldsymbol{W}_1^T \mathbf{x}))^2\, \rho\, d\mathbf{x} \right)^{\frac{1}{2}}. \tag{4.38}$$

Theorem 4.4 bounds the first summand. By the tower property and Assumption 1, the second summand satisfies

$$\int (\hat{g}(\boldsymbol{W}_1^T \mathbf{x}) - \mathscr{R}(\boldsymbol{W}_1^T \mathbf{x}))^2\, \rho\, d\mathbf{x} = \int \left(\int (\hat{g}(\mathbf{y}) - \mathscr{R}(\mathbf{y}))^2\, \pi_{Z|Y}\, d\mathbf{z} \right) \pi_Y\, d\mathbf{y} \leq (C_2 \delta)^2. \tag{4.39}$$

Take the square root to complete the proof. $\square$

The next equation summarizes the three levels of approximation:

$$f(\mathbf{x}) \approx g(\boldsymbol{W}_1^T\mathbf{x}) \approx \hat{g}(\boldsymbol{W}_1^T\mathbf{x}) \approx \mathscr{R}(\boldsymbol{W}_1^T\mathbf{x}). \tag{4.40}$$

The conditional expectation g is defined in (4.15), its Monte Carlo approximation $\hat{g}$ is defined in (4.26), and the response surface $\mathscr{R}$ is defined in (4.34). The respective error estimates are given in Theorems 4.3, 4.4, and 4.5.

4.2.2 ▪ Using estimated eigenvectors

Until now, we have used the exact eigenvectors $\boldsymbol{W}$. In practice we have estimates $\hat{\boldsymbol{W}}$ from the procedure in section 3.4. The approximate eigenvectors define a rotation of the domain of f that is slightly different from the rotation defined by the true eigenvectors. We can construct the same approximations and response surfaces using this slightly different rotation. In this section we examine the estimated eigenvectors' effects on the approximation of f. We assume the following characterization of the estimates.

Assumption 2. *Let $\boldsymbol{W}$ and its estimate $\hat{\boldsymbol{W}}$ be comparably partitioned,*

$$\boldsymbol{W} = \begin{bmatrix} \boldsymbol{W}_1 & \boldsymbol{W}_2 \end{bmatrix}, \quad \hat{\boldsymbol{W}} = \begin{bmatrix} \hat{\boldsymbol{W}}_1 & \hat{\boldsymbol{W}}_2 \end{bmatrix}, \tag{4.41}$$

where $\boldsymbol{W}_1$ and $\hat{\boldsymbol{W}}_1$ contain the first $n < m$ columns of $\boldsymbol{W}$ and $\hat{\boldsymbol{W}}$, respectively. Then there is an $\varepsilon > 0$ such that

$$\mathrm{dist}(\mathrm{ran}(\boldsymbol{W}_1), \mathrm{ran}(\hat{\boldsymbol{W}}_1)) \leq \varepsilon, \tag{4.42}$$

where $\mathrm{dist}(\mathrm{ran}(\boldsymbol{W}_1), \mathrm{ran}(\hat{\boldsymbol{W}}_1))$ is defined in (3.49).

Compare Assumption 2 to Corollary 3.10 from the previous chapter, and note that the distance between the true and the estimated subspaces depends on the corresponding spectral gap.

The estimate $\hat{\boldsymbol{W}}$ defines approximate active variables $\hat{\mathbf{y}} = \hat{\boldsymbol{W}}_1^T\mathbf{x}$ and inactive variables $\hat{\mathbf{z}} = \hat{\boldsymbol{W}}_2^T\mathbf{x}$, which have comparable joint, marginal, and conditional densities:

$$\hat{\pi}(\hat{\mathbf{y}},\hat{\mathbf{z}}) = \rho(\hat{\boldsymbol{W}}_1\hat{\mathbf{y}} + \hat{\boldsymbol{W}}_2\hat{\mathbf{z}}), \tag{4.43}$$

$$\hat{\pi}_{\hat{Y}}(\hat{\mathbf{y}}) = \int \hat{\pi}(\hat{\mathbf{y}},\hat{\mathbf{z}})\,d\hat{\mathbf{z}}, \qquad \hat{\pi}_{\hat{Z}}(\hat{\mathbf{z}}) = \int \hat{\pi}(\hat{\mathbf{y}},\hat{\mathbf{z}})\,d\hat{\mathbf{y}}, \tag{4.44}$$

$$\hat{\pi}_{\hat{Z}|\hat{Y}}(\hat{\mathbf{z}}|\hat{\mathbf{y}}) = \frac{\hat{\pi}(\hat{\mathbf{y}},\hat{\mathbf{z}})}{\hat{\pi}_{\hat{Y}}(\hat{\mathbf{y}})}, \qquad \hat{\pi}_{\hat{Y}|\hat{Z}}(\hat{\mathbf{y}}|\hat{\mathbf{z}}) = \frac{\hat{\pi}(\hat{\mathbf{y}},\hat{\mathbf{z}})}{\hat{\pi}_{\hat{Z}}(\hat{\mathbf{z}})}. \tag{4.45}$$

The domain of the perturbed approximations is

$$\hat{\mathscr{Y}} = \left\{ \hat{\mathbf{y}} \in \mathbb{R}^n : \hat{\mathbf{y}} = \hat{\boldsymbol{W}}_1^T\mathbf{x},\, \mathbf{x} \in \mathscr{X} \right\}. \tag{4.46}$$

We construct the same sequence of approximations of f using the estimated active variables. Denote the perturbed versions of the approximations with a subscript ε. The conditional expectation of f given $\hat{\mathbf{y}}$ becomes

$$f(\mathbf{x}) \approx g_\varepsilon(\hat{\boldsymbol{W}}_1^T\mathbf{x}), \tag{4.47}$$

where

$$g_\varepsilon(\hat{\mathbf{y}}) = \int f(\hat{\mathbf{W}}_1\hat{\mathbf{y}} + \hat{\mathbf{W}}_2\hat{\mathbf{z}})\,\hat{\pi}_{\hat{Z}|\hat{Y}}\,d\hat{\mathbf{z}}. \tag{4.48}$$

Then we have the following error estimate.

Theorem 4.6. *The root-mean-squared error in the conditional expectation g_ε using the estimated active variables satisfies*

$$\left(\int (f(\mathbf{x}) - g_\varepsilon(\hat{\mathbf{W}}_1^T\mathbf{x}))^2\,\rho\,d\mathbf{x}\right)^{\frac{1}{2}} \leq C_1\left(\varepsilon\,(\lambda_1 + \cdots + \lambda_n)^{\frac{1}{2}} + (\lambda_{n+1} + \cdots + \lambda_m)^{\frac{1}{2}}\right), \tag{4.49}$$

where C_1 is from Theorem 4.3.

Proof. Following the same reasoning as in the proof of Theorem 4.3 and using the Poincaré inequalities from Lemmas 4.1 and 4.2, we get

$$\int (f(\mathbf{x}) - g_\varepsilon(\hat{\mathbf{W}}_1^T\mathbf{x}))^2\,\rho\,d\mathbf{x} \leq C_1^2 \int (\nabla_{\hat{z}}f(\mathbf{x}))^T(\nabla_{\hat{z}}f(\mathbf{x}))\,\rho\,d\mathbf{x}. \tag{4.50}$$

Using the chain rule, $\nabla_{\hat{z}}f = \hat{\mathbf{W}}_2^T\mathbf{W}_1\nabla_{\mathbf{y}}f + \hat{\mathbf{W}}_2^T\mathbf{W}_2\nabla_{\mathbf{z}}f$. Note that Assumption 2 implies $\|\mathbf{W}_1^T\hat{\mathbf{W}}_2\| \leq \varepsilon$, and $\|\mathbf{W}_2^T\hat{\mathbf{W}}_2\| \leq 1$. Then

$$\nabla_{\hat{z}}f^T\nabla_{\hat{z}}f = \left(\hat{\mathbf{W}}_2^T\mathbf{W}_1\nabla_{\mathbf{y}}f + \hat{\mathbf{W}}_2^T\mathbf{W}_2\nabla_{\mathbf{z}}f\right)^T\left(\hat{\mathbf{W}}_2^T\mathbf{W}_1\nabla_{\mathbf{y}}f + \hat{\mathbf{W}}_2^T\mathbf{W}_2\nabla_{\mathbf{z}}f\right) \tag{4.51}$$

$$= \nabla_{\mathbf{y}}f^T\mathbf{W}_1^T\hat{\mathbf{W}}_2\hat{\mathbf{W}}_2^T\mathbf{W}_1\nabla_{\mathbf{y}}f + 2\nabla_{\mathbf{y}}f^T\mathbf{W}_1^T\hat{\mathbf{W}}_2\hat{\mathbf{W}}_2^T\mathbf{W}_2\nabla_{\mathbf{z}}f \tag{4.52}$$

$$+ \nabla_{\mathbf{z}}f^T\mathbf{W}_2^T\hat{\mathbf{W}}_2\hat{\mathbf{W}}_2^T\mathbf{W}_2\nabla_{\mathbf{z}}f \tag{4.53}$$

$$\leq \varepsilon^2\nabla_{\mathbf{y}}f^T\nabla_{\mathbf{y}}f + 2\varepsilon\nabla_{\mathbf{y}}f^T\nabla_{\mathbf{z}}f + \nabla_{\mathbf{z}}f^T\nabla_{\mathbf{z}}f. \tag{4.54}$$

Averaging both sides and applying the Cauchy–Schwarz inequality,

$$\int (\nabla_{\hat{z}}f(\mathbf{x}))^T(\nabla_{\hat{z}}f(\mathbf{x}))\,\rho\,d\mathbf{x} \tag{4.55}$$

$$\leq \int (\nabla_{\mathbf{z}}f(\mathbf{x}))^T(\nabla_{\mathbf{z}}f(\mathbf{x}))\,\rho\,d\mathbf{x} + 2\varepsilon \int (\nabla_{\mathbf{z}}f(\mathbf{x}))^T(\nabla_{\mathbf{y}}f(\mathbf{x}))\,\rho\,d\mathbf{x} \tag{4.56}$$

$$+ \varepsilon^2 \int (\nabla_{\mathbf{y}}f(\mathbf{x}))^T(\nabla_{\mathbf{y}}f(\mathbf{x}))\,\rho\,d\mathbf{x} \tag{4.57}$$

$$\leq \left(\left(\int (\nabla_{\mathbf{z}}f(\mathbf{x}))^T(\nabla_{\mathbf{z}}f(\mathbf{x}))\,\rho\,d\mathbf{x}\right)^{\frac{1}{2}} + \varepsilon\left(\int (\nabla_{\mathbf{y}}f(\mathbf{x}))^T(\nabla_{\mathbf{y}}f(\mathbf{x}))\,\rho\,d\mathbf{x}\right)^{\frac{1}{2}}\right)^2 \tag{4.58}$$

$$\leq \left(\varepsilon\,(\lambda_1 + \cdots + \lambda_n)^{\frac{1}{2}} + (\lambda_{n+1} + \cdots + \lambda_m)^{\frac{1}{2}}\right)^2. \tag{4.59}$$

The last line follows from Lemma 3.2. $\qquad\square$

The eigenvalues $\lambda_1, \dots, \lambda_n$ associated with the true active subspace contribute to the error estimate. This contribution persists in error estimates for the Monte Carlo approximation and the response surface using the estimated eigenvectors.

Define the Monte Carlo estimate as

$$\hat{g}_\varepsilon(\hat{\mathbf{y}}) = \frac{1}{N} \sum_{i=1}^{N} f(\hat{\mathbf{W}}_1 \hat{\mathbf{y}} + \hat{\mathbf{W}}_2 \hat{\mathbf{z}}_i), \tag{4.60}$$

where the $\hat{\mathbf{z}}_i$'s are drawn independently from the conditional density $\hat{\pi}_{\hat{Z}|\hat{Y}}$. Then we have the following error estimate, whose derivation follows the proof of Theorem 4.4 using the estimated active variables and the reasoning from the proof of Theorem 4.6.

Theorem 4.7. *The root-mean-squared error in the Monte Carlo approximation $\hat{g}_\varepsilon$ using the estimated eigenvectors $\hat{\mathbf{W}}_1$ satisfies*

$$\left(\int (f(\mathbf{x}) - \hat{g}_\varepsilon(\hat{\mathbf{W}}_1^T \mathbf{x}))^2 \, \rho \, d\mathbf{x} \right)^{\frac{1}{2}} \tag{4.61}$$
$$\leq C_1 \left(1 + N^{-\frac{1}{2}} \right) \left(\varepsilon \, (\lambda_1 + \cdots + \lambda_n)^{\frac{1}{2}} + (\lambda_{n+1} + \cdots + \lambda_m)^{\frac{1}{2}} \right),$$

where C_1 and N are the quantities from Theorem 4.4.

The response surface approximation using the estimated eigenvectors is

$$\hat{g}_\varepsilon(\hat{\mathbf{y}}) \approx \mathscr{R}_\varepsilon(\hat{\mathbf{y}}; \hat{g}_{\varepsilon,1}, \dots, \hat{g}_{\varepsilon,P}) \tag{4.62}$$

for a chosen response surface method $\mathscr{R}_\varepsilon$. The subscript ε indicates that the response surface is built on the estimated active variables. The $\hat{g}_{\varepsilon,k}$'s are evaluations of the Monte Carlo estimates $\hat{g}_\varepsilon$ at the fixed design points $\hat{\mathbf{y}}_k \in \hat{\mathscr{Y}}$. We have the following error estimate; again, its derivation follows the proof of Theorem 4.5 using the reasoning from the proof of Theorem 4.6.

Theorem 4.8. *Under the assumptions of Theorem 4.5 and assuming that an error bound comparable to (4.36) holds for $\hat{g}_\varepsilon$ and $\mathscr{R}_\varepsilon$, the root-mean-squared error in the response surface approximation $\mathscr{R}_\varepsilon$ satisfies*

$$\left(\int (f(\mathbf{x}) - \mathscr{R}_\varepsilon(\hat{\mathbf{W}}_1^T \mathbf{x}))^2 \, \rho \, d\mathbf{x} \right)^{\frac{1}{2}} \tag{4.63}$$
$$\leq C_1 \left(1 + N^{-\frac{1}{2}} \right) \left(\varepsilon \, (\lambda_1 + \cdots + \lambda_n)^{\frac{1}{2}} + (\lambda_{n+1} + \cdots + \lambda_m)^{\frac{1}{2}} \right) + C_2 \, \delta,$$

where C_1, N, C_2, and δ are the quantities from Theorem 4.5.

We summarize the three levels of approximation using the estimated eigenvectors

$$f(\mathbf{x}) \approx g_\varepsilon(\hat{\mathbf{W}}_1^T \mathbf{x}) \approx \hat{g}_\varepsilon(\hat{\mathbf{W}}_1^T \mathbf{x}) \approx \mathscr{R}_\varepsilon(\hat{\mathbf{W}}_1^T \mathbf{x}). \tag{4.64}$$

The conditional expectation g_ε is defined in (4.48), its Monte Carlo approximation $\hat{g}_\varepsilon$ is defined in (4.60), and the response surface $\mathscr{R}_\varepsilon$ is defined in (4.62). The respective error estimates are given in Theorems 4.6, 4.7, and 4.8.

4.2.3 ▪ Example: The parameterized PDE

In previous work [28, section 5], we constructed a kriging surface (also known as a Gaussian process regression) on the one-dimensional active subspace in the parameterized PDE example from section 3.4.2. We show the results of that study here.

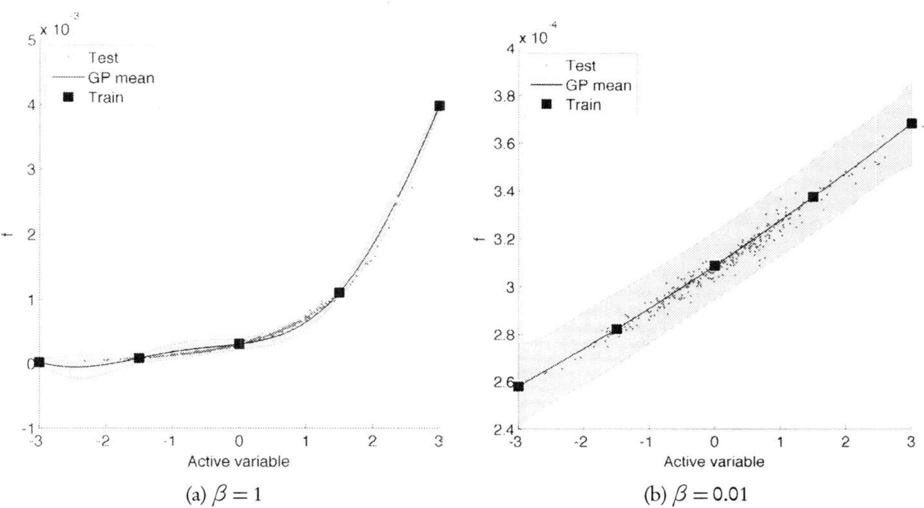

(a) $\beta = 1$ (b) $\beta = 0.01$

Figure 4.2. *Sufficient summary plots of the quantity of interest from the parameterized PDE against the first active variable. The univariate kriging surfaces overlay the sufficient summary plots; the black solid line is the mean prediction, and the gray shaded region shows the two-standard deviation prediction interval. Modified from* [28].

We compare the kriging surface constructed on the one-dimensional $\hat{\mathcal{Y}}$ with a kriging surface on the full 100-dimensional domain $\mathcal{X}$. The one-dimensional kriging surface includes a quadratic mean term, and the 100-dimensional kriging surface uses a linear mean term. More modeling choices become feasible with fewer dimensions, such as using a quadratic mean instead of a linear mean. Both surfaces use a squared exponential correlation kernel, and we fit the correlation parameters with maximum likelihood as described by Rasmussen and Williams [106].

The cost of computing the gradient $\nabla_{\mathbf{x}} f(\mathbf{x})$ via the adjoint is roughly twice the cost of computing the function f for a particular $\mathbf{x}$. When evaluating the Monte Carlo estimate $\hat{g}_\varepsilon$ of the conditional expectation, we choose $N = 1$ sample per design point. Thus, the cost of discovering and exploiting the active subspace is roughly $3M + P$ function evaluations, where M is the number of gradient samples, and P is the number of design points on $\hat{\mathcal{Y}}$. In this case $M = 300$, which corresponds to a multiplier $\alpha = 10$, $k = 6$ eigenvalues sought, and $m = 100$ dimensions from Algorithm 3.3. We build the kriging surface with $P = 5$ design points.

Figure 4.2 shows the sufficient summary plots using one active variable for both long and short correlation lengths. The solid line is the mean kriging prediction, and the gray shaded region is the two-standard deviation prediction interval. The dots show the 300 evaluations of f computed while sampling the gradient to discover the subspace. Figure 4.2(a) shows the approximation for the long correlation length. Notice how the function evaluations cluster tightly around the mean prediction. This confirms that the active subspace has found an angle from which to view the high-dimensional data to uncover its one-dimensional character. Figure 4.2(b) shows the same plot for the shorter correlation length; it suggests similar conclusions.

For a fair comparison, we build a kriging surface on the 100-dimensional space $\mathcal{X}$ using $3M + P = 905$ function evaluations, which are enough samples to fit the kriging surface

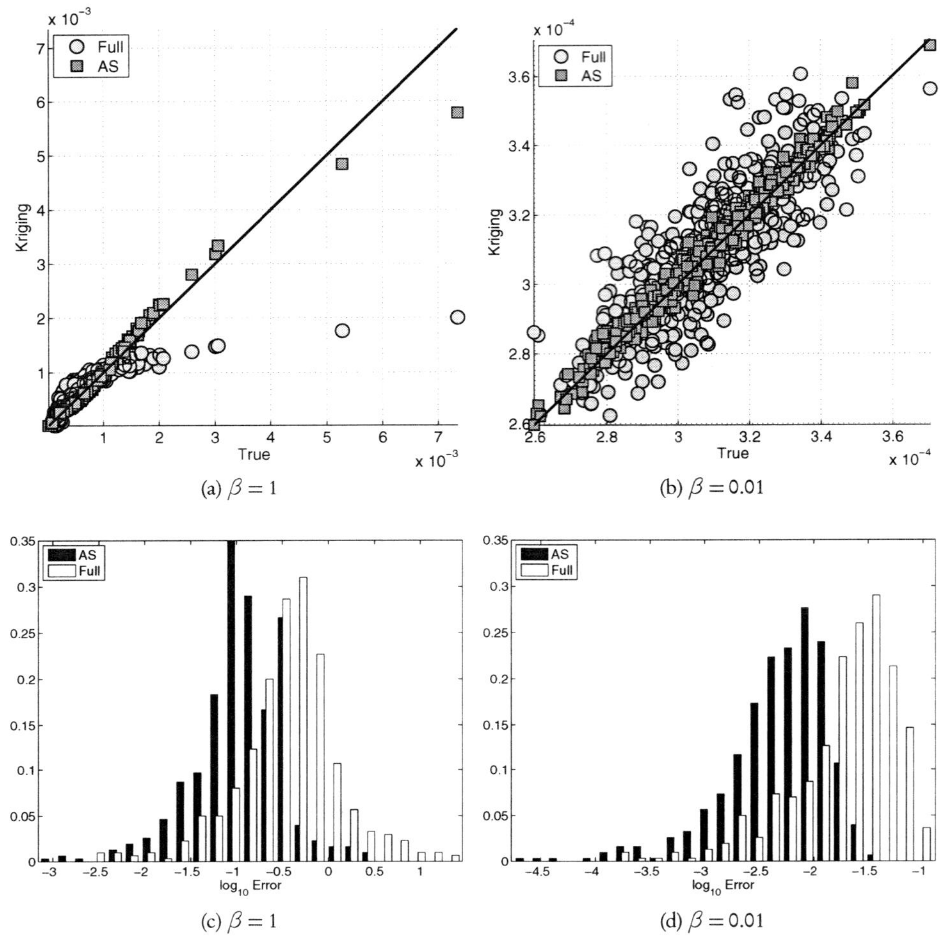

Figure 4.3. *(a) and (b) plot the true value of the testing data on the horizontal axis against the kriging surface approximation on the vertical axis for both the surface constructed on one active variable and the surface constructed on all* 100 *variables. The relatively small spread about the diagonal for the active variable construction indicates that it is outperforming the approximation on all* 100 *variables. This is confirmed by the histograms of the log of the relative error in the testing data. In the legends, "AS" is the kriging surface using the active subspace, and "Full" is the kriging surface on all* 100 *variables. Images (c) and (d) are modified from* [28].

with a linear mean term, but not enough for a quadratic mean term in 100 variables. We evaluate f at 500 additional random points to create an independent testing set. Figure 4.3 shows two representations of the error to compare the two kriging approximations. (a) and (b) plot the true function value on the horizontal axis against its kriging approximation on the vertical axis; the closer to the diagonal, the better. The surface constructed on the active variable performs better than the surface constructed on all 100 variables. This is confirmed by (c) and (d), which show histograms of the log of the relative error. The masses of the error histograms for the active variable construction are to the left of the error histogram for the full construction, which indicates better performance. The kriging surface on the active variable is able to focus its effort on the active direction in which f changes, and it does not need samples along the space orthogonal to the active direction.

4.2.4 ▪ Remarks

We conclude this section with a few remarks. First, throughout Chapter 3 we suggested that one should choose the active subspace's dimension n by looking for *gaps* between the eigenvalues; a relatively large gap implies that the subspace is better approximated. However, the error estimates in this chapter depend on the magnitude of the eigenvalues. So there is a subtle balance in choosing the dimension of the active subspace for response surfaces. The dimension n should be chosen so that $\lambda_{n+1}, \ldots, \lambda_m$ are reasonably small. But the estimated subspace must be accurate enough to keep the ε small in Theorems 4.6, 4.7, and 4.8. Otherwise errors include quantities close to the larger eigenvalues.

Second, the conditional expectation $g(\mathbf{y})$ is a continuous function of $\mathbf{y}$, but it may not be differentiable. For example, if the domain $\mathcal{X}$ is a hypercube, then projecting the domain and averaging the function over the inactive variables can create *kinks* in $g(\mathbf{y})$. This is a result of the sharp corners of the domain.

Third, the proofs of Theorems 4.4 and 4.7 assume that one can draw $\mathbf{z}_i$ and $\hat{\mathbf{z}}_i$ independently from the conditional densities $\pi_{Z|Y}$ and $\pi_{\hat{Z}|\hat{Y}}$, respectively. If ρ is a standard Gaussian, then drawing independent samples is straightforward. For other cases in practice—including the uniform density on a hypercube—we may need to use a Metropolis–Hastings method [27] to sample from $\pi_{Z|Y}$. But these samples are correlated, and the error bound in (4.31) does not strictly apply [23].

Fourth, there is substantial freedom in choosing both the design sites on the domain $\mathcal{Y}$ and the response surface. This was intentional. The approximation details for a specific application require many more practical considerations than we can address here. In prior work [28], we used kriging primarily because of (i) the natural fit of the computed λ_i to the correlation length parameters and training data noise model, (ii) its flexibility with scattered design sites, and (iii) its lack of strict interpolation. However, many other options for approximation are possible including global polynomials, regression splines, or finite element approximations. As Koehler and Owen discuss [76], we could also use the correlation information to construct efficient designs on the low-dimensional domain $\mathcal{Y}$, i.e., designs that satisfy optimality criteria such as maximum entropy.

4.3 ▪ Integration

We turn to the next parameter study, namely, what is the simulation output's average over all possible input values? This question is phrased precisely as an integral,

$$I = \int f(\mathbf{x})\rho\,d\mathbf{x}. \tag{4.65}$$

One can similarly compute moments or quantiles of f by replacing the integrand with various functionals of f. Numerical integration (or *quadrature*) rules approximate the integral with a weighted sum of function evaluations,

$$I \approx \sum_{j=1}^{M} f(\mathbf{x}_j)\,w_j, \tag{4.66}$$

where the $\mathbf{x}_j$'s and w_j's are the nodes and weights, respectively, of the quadrature rule. The classical text on numerical integration methods is Davis and Rabinowitz [43], and a survey by Hickernell [67] provides a good launching pad for further study in modern methods for numerically integrating high-dimensional integrands. Sparse grid-based methods have also risen to prominence for efficiently integrating certain classes of functions in high dimensions [19]. O'Hagan offers an alternative Bayesian perspective on numerical integration [97].

A common way to build multivariate quadrature rules is by taking tensor products of well-studied univariate rules. But this construction is impractical in high dimensions because the number of function evaluations needed to compute the numerical approximation scales exponentially with the dimension. We want to exploit the active subspace to reduce the dimension of the integrand and, consequently, the work needed to estimate the integral. The inactive subspace identifies directions in the input space where the integrand is relatively flat. Averaging constants is easy: the average of a constant is the constant. If the active subspace identifies a few directions along which f varies, then we can focus the effort of a quadrature rule along those directions.

The ideas in this section are not nearly as complete as the analysis of response surfaces; these integration ideas are emerging. We have some general ideas and some inklings of what to do with special cases. But the theory and methods are still open for development.

The essential connection between the integral and the active subspace is the tower property of the conditional expectation [126],

$$\int f(\mathbf{x})\rho\,d\mathbf{x} = \int \left(\int f(\mathbf{W}_1\mathbf{y} + \mathbf{W}_2\mathbf{z})\,\pi_{Z|Y}\,d\mathbf{z} \right) \pi_Y\,d\mathbf{y} = \int g(\mathbf{y})\,\pi_Y\,d\mathbf{y}, \qquad (4.67)$$

where g is the conditional expectation of f given $\mathbf{y}$ defined in (4.15). This relation suggests that we can use a quadrature rule on only the $n < m$ active variables $\mathbf{y}$,

$$I \approx I_N = \sum_{k=1}^{N} g(\mathbf{y}_k)\,w_k, \qquad (4.68)$$

for some nodes $\mathbf{y}_k \in \mathcal{Y}$ from (4.6) and weights w_k related to the marginal density π_Y. This reasoning raises the following two questions.

1. How should we choose the nodes and weights?

2. How should we approximate the conditional expectation g at each node?

To address the second question, we have proposed a Monte Carlo method which we define in (4.26). If the nodes and weights are given, then we can approximate

$$I_N \approx \hat{I}_N = \sum_{k=1}^{N} \hat{g}(\mathbf{y}_k)\,w_k. \qquad (4.69)$$

It is difficult to estimate the error in $\hat{g}(\mathbf{y}_k)$. However, recall from the proof of Theorem 4.4 that we can bound the average conditional variance. The conditional variance $\sigma_\mathbf{y}^2$ is

$$\sigma_\mathbf{y}^2 = \int (f(\mathbf{W}_1\mathbf{y} + \mathbf{W}_2\mathbf{z}) - g(\mathbf{y}))^2\,\pi_{Z|Y}\,d\mathbf{z}. \qquad (4.70)$$

Averaging this quantity over $\mathbf{y}$, we get

$$\int \sigma_\mathbf{y}^2\,\pi_Y\,d\mathbf{y} \leq C_1^2 \left(\lambda_{n+1} + \cdots + \lambda_m \right). \qquad (4.71)$$

If $\lambda_{n+1} = \cdots = \lambda_m = 0$, then (4.71) implies that $\hat{g}(\mathbf{y}_k) = g(\mathbf{y}_k)$ for each $\mathbf{y}_k$, and $\hat{I}_N = I_N$; the only error in approximating the integral is the error in the integration rule on the active variables $\mathbf{y}$. But this is a very special case. In practice, the last $m - n$ eigenvalues may be small but not zero. We hypothesize that discrepancies between $g(\mathbf{y}_k)$ and $\hat{g}(\mathbf{y}_k)$ cancel out in the summation, yielding good estimates of the discretized integral I_N; this is work in progress.

Another option is to use the central limit theorem to derive confidence intervals on $\hat{I}_N$. If we can draw $\mathbf{z}_i$ independently from the conditional density $\pi_{Z|Y}$, then the random variables $\hat{g}(\mathbf{y}_k)$ are approximately Gaussian with mean $g(\mathbf{y}_k)$ and variance $\sigma_{\mathbf{y}_k}^2/M$, where M is the number of independent samples from $\pi_{Z|Y}$. The linear combination of approximately Gaussian random variables $\hat{I}_N$ is approximately Gaussian for a fixed N. We get plug-in estimates for its variance and, consequently, estimates of the confidence intervals. In practice, we may need to draw correlated samples from $\pi_{Z|Y}$, since it may only be known up to a constant factor. Confidence and error estimates should be adjusted accordingly [23].

Next we attend to the nodes $\mathbf{y}_k$ and weights w_k from (4.68). We focus on two commonly encountered special cases: (i) the *Gaussian case*, where f's domain is $\mathbb{R}^m$ and the probability density ρ is a standard Gaussian density, and (ii) the *uniform case*, where f's domain is the hypercube $[-1, 1]^m$ and ρ is uniform in $[-1, 1]^m$ and zero elsewhere.

4.3.1 ▪ The Gaussian case

For the Gaussian case, we assume the domain $\mathscr{X}$ of the simulation $f(\mathbf{x})$ is all of $\mathbb{R}^m$. Then the range of the active variables is $\mathbb{R}^n$, and the range of the inactive variables is $\mathbb{R}^{m-n}$. The probability density function $\rho(\mathbf{x})$ is a Gaussian with zero mean and identity covariance. Then the joint density $\pi(\mathbf{y}, \mathbf{z})$, the marginal densities $\pi_Y(\mathbf{y})$, $\pi_Z(\mathbf{z})$, and the conditional densities $\pi_{Z|Y}(\mathbf{z}|\mathbf{y})$, $\pi_{Y|Z}(\mathbf{y}|\mathbf{z})$ are all Gaussian with zero mean and identity covariance. In this case, drawing independent samples from $\pi_{Z|Y}$ for the Monte Carlo estimates $\hat{g}(\mathbf{y}_k)$ of the conditional expectation $g(\mathbf{y}_k)$ is straightforward.

The quadrature nodes and weights can be treated like the sites for a randomized design of experiments, where the weights are $w_k = 1/N$. Methods for generating designs on the cube with a uniform density translate to designs on unbounded domains with a Gaussian density through the error function; see, e.g., section 3.4 of Caflisch's survey paper [20]. When the number of active variables is small enough, one can consider deterministic integration rules for infinite domains and Gaussian weight functions, such as the Genz–Keister rules [57] or sparse grid rules for unbounded domains [96]. If there is only one active variable, then several options exist for the one-dimensional, infinite interval [43].

4.3.2 ▪ The uniform case

In the uniform case, the domain is the hypercube $[-1, 1]^m$ equipped with a uniform weight function that is zero outside the hypercube. We can also write the domain as

$$\mathscr{X} = \{\mathbf{x} \in \mathbb{R}^m : -1 \le \mathbf{x} \le 1\}, \tag{4.72}$$

where $\mathbf{1}$ is a vector of ones. This case is considerably more complicated than the Gaussian case. The range of the active variables $\mathscr{Y}$ from (4.6) is the image of a hypercube under a linear transformation from $\mathbb{R}^m$ to $\mathbb{R}^n$. The 2^m vertices from the hypercube get transformed, and the convex hull of these points defines a polytope in $\mathbb{R}^n$; this object is called a *zonotope* [53]. There are algorithms for identifying the zonotope vertices that are polynomial in n, and these vertices define the zonotope boundary.

The marginal density π_Y is

$$\pi_Y(\mathbf{y}) = \int \pi(\mathbf{y}, \mathbf{z}) \, d\mathbf{z} = 2^{-m} \int_{\mathscr{S}_\mathbf{y}} d\mathbf{z} = 2^{-m} \operatorname{Vol}(\mathscr{S}_\mathbf{y}), \tag{4.73}$$

where the set $\mathscr{S}_\mathbf{y}$ is the polytope

$$\mathscr{S}_\mathbf{y} = \{\mathbf{z} \in \mathbb{R}^{m-n} : -1 - W_1\mathbf{y} \le W_2\mathbf{z} \le 1 - W_1\mathbf{y}\}, \tag{4.74}$$

and $\operatorname{Vol}(\cdot)$ returns the volume of the argument. Computing a polytope's volume in several dimensions is generally difficult. Several algorithms are available for approximating the volume [16], including randomized methods [86]. Thorough analysis of the integration error for I_N from (4.68) should include errors in approximating the marginal density, since the quadrature weights w_k may depend on the marginal density.

Given the boundary of the zonotope and the marginal density π_Y, one could, in principle, construct quasi–Monte Carlo sequences to approximate the integral [20]. We are currently pursuing this idea.

If the number n of active variables is small enough—say, less than four—we could construct a Voronoi tessellation of the zonotope $\mathscr{Y}$ [48]. If $\{V_k\}$ are the cells of the tessellation, then each quadrature node $\mathbf{y}_k$ would be the centroid of the cell V_k. The weights would be the integral of the marginal density over the cell,

$$w_k = \int_{V_k} \pi_Y(\mathbf{y}) \, d\mathbf{y}. \tag{4.75}$$

Such rules are optimal for certain classes of functions [48, section 2.2]. We could approximate the integrals (4.75) with sampling, since drawing $\mathbf{x}$ uniformly from the hypercube and applying the map $\mathbf{y} = W_1^T \mathbf{x}$ is cheap and simple. In other words, we could draw an extreme number of samples of $\mathbf{y}$ and divide the number of samples that land in V_k by the total number of samples to estimate w_k.

4.4 ▪ Optimization

Minimizing the simulation's prediction over all possible inputs is stated simply in our notation: find $\mathbf{x}^* \in \mathscr{X}$ such that $f(\mathbf{x}^*) \le f(\mathbf{x})$ for all $\mathbf{x} \in \mathscr{X}$. We can also write this as

$$\underset{\mathbf{x}}{\text{minimize}} \, f(\mathbf{x}) \text{ subject to } \mathbf{x} \in \mathscr{X}. \tag{4.76}$$

Standard formulations of optimization may include additional constraints on the variables, such as linear or nonlinear equality or inequality constraints. We assume that any such constraints will be included in the definition of $f(\mathbf{x})$ with established techniques, e.g., barrier formulations [95]. We also assume that the minimizer exists, but it may not be unique. In fact, the existence of an active subspace suggests that there may be subsets of the input space where f is very close to its minimum.

Optimization methods are replete with subspaces—usually derived from the eigenvectors of f's Hessian. The Hessian is a local quantity, and the subspaces of interest change at each iteration. In contrast, the active subspace is a global quantity that describes the directions along which input perturbations change f more, on average. If the Hessian or Hessian approximation is expensive to evaluate, and if we have an idea of f's active subspace, then we would wish to exploit the active subspace to improve or accelerate the optimization method.

We rewrite (4.76) in terms of active and inactive variables as

$$\underset{\mathbf{y},\mathbf{z}}{\text{minimize}}\, f(W_1\mathbf{y} + W_2\mathbf{z}) \text{ subject to } W_1\mathbf{y} + W_2\mathbf{z} \in \mathcal{X}. \tag{4.77}$$

If the eigenvalues associated with the inactive subspace are zero, $\lambda_{n+1} = \cdots = \lambda_m = 0$, and if the domain $\mathcal{X}$ is all $\mathbb{R}^m$, then we can set $\mathbf{z} = 0$ and restrict our attention to the subset of $\mathbb{R}^m$ spanned by the eigenvectors W_1. In other words, we can rewrite (4.77) as

$$\underset{\mathbf{y}}{\text{minimize}}\, f(W_1\mathbf{y}). \tag{4.78}$$

The situation is more complicated if $\mathcal{X}$ is bounded. Suppose we solve

$$\underset{\mathbf{y}}{\text{minimize}}\, f(W_1\mathbf{y}) \text{ subject to } W_1\mathbf{y} \in \mathcal{X}. \tag{4.79}$$

Let $\mathbf{y}^*$ be the minimizer of (4.79). If $\mathcal{X}$ is a hypercube, then the constraint $W_1\mathbf{y} \in \mathcal{X}$ can be expressed as a set of linear inequality constraints on $\mathbf{y}$. If the minimizer $\mathbf{y}^*$ is in the interior of $\mathcal{X}$, then we are done. But if it lives on the boundary, $W_1\mathbf{y}^* \in \partial\mathcal{X}$, then we may have more to do. In particular, suppose the vector $\mathbf{s}$ is such that

$$W_1\mathbf{y}^* + \mathbf{s} \in \partial\mathcal{X}, \qquad W_1^T\mathbf{s} \neq 0. \tag{4.80}$$

Then moving to the point $\mathbf{y}^* + W_1^T\mathbf{s}$ should change the value of f. In general, it may be possible to move along the boundary of the domain in directions that have nonzero components in the range of the eigenvectors W_1. Such moves could further reduce f. This emphasizes the subtle relationship between the active subspace and the space of inputs for f. The active subspace describes a set of directions along which f changes. It does not necessarily restrict our attention to a subset of the input space.

The more interesting case is when the domain $\mathcal{X}$ is bounded and the eigenvalues associated with the inactive subspace are not exactly zero. For concreteness, assume $\mathcal{X}$ is the hypercube $[-1, 1]^m$. What conditions should f satisfy to ensure that knowing the active subspace leads to advantages in optimization? And how can we modify the algorithms to exploit the active subspace? Consider the following construction for a function $g = g(\mathbf{y})$ of the active variables that may be useful for optimization:

$$g(\mathbf{y}) = \left\{ \begin{array}{ll} \underset{\mathbf{z}}{\text{minimum}} & f(W_1\mathbf{y} + W_2\mathbf{z}), \\ \text{subject to} & -1 - W_1\mathbf{y} \leq W_2\mathbf{z} \leq 1 - W_1\mathbf{y}. \end{array} \right. \tag{4.81}$$

In other words, for each $\mathbf{y} \in \mathcal{Y}$ from (4.6), we set g to be the minimum of f with $\mathbf{y}$ fixed and $\mathbf{z}$ varying such that $\mathbf{x} = W_1\mathbf{y} + W_2\mathbf{z}$ remains in the domain of f. We might impose additional regularizing constraints on $\mathbf{z}$ to ensure that the minimizer is unique.

With the construction in (4.81), we could try to minimize g as

$$\underset{\mathbf{y}}{\text{minimize}}\, g(\mathbf{y}) \text{ subject to } \mathbf{y} \in \mathcal{Y}. \tag{4.82}$$

This is an optimization problem in $n < m$ variables. If $\mathbf{y}^*$ is the minimizer of (4.82), then we hope that f's minimizer might reside in the set of $\mathbf{x}$'s such that $\mathbf{y}^* = W_1^T\mathbf{x}$. We imagine that the active subspace may help with the first few iterations of a general method for minimizing f, or it might help to find a good starting point.

There are several potential issues with this approach. First, the gradient of g with respect to $\mathbf{y}$ may be discontinuous. A natural way to define the gradient $\nabla_\mathbf{y} g(\mathbf{y})$ might be

$$\nabla_\mathbf{y} g(\mathbf{y}) = W_1^T \nabla_\mathbf{x} f(\mathbf{x}^*), \qquad \mathbf{x}^* = W_1 \mathbf{y} + W_2 \mathbf{z}^*, \qquad (4.83)$$

where $\mathbf{z}^*$ minimizes (4.81). If the minimizer is not unique, then the choice of gradient is ambiguous. In fact, some gradient directions might yield ascent directions for g. A 2-norm (i.e., Tikhonov) regularization on $\mathbf{z}$ to make the minimizer unique may inadvertently produce ascent directions when other regularizations might yield descent directions. We are currently pursuing the idea of how to choose a regularization on (4.81) that admits a satisfying definition of the gradient $\nabla_\mathbf{y} g$.

Second, the minimum of $g(\mathbf{y})$ may not be the minimum of $f(\mathbf{x})$. We are working to find the conditions that guarantee that these two minima are the same. The function f has more variables to change to seek a minimum, so we expect its minimum to be smaller than g's minimum. But if f's active subspace is dominant, then g's minimum might be close to f's. The minimizer of g might produce a good starting point for Newton's method; choosing the starting point is challenging in high dimensions when f is not convex.

Third, g's construction in (4.81) does not reduce computation. Minimizing over $\mathbf{z}$ at each iteration of minimizing over $\mathbf{y}$ is at least as expensive as minimizing over $\mathbf{x}$. However, we might save computation if we can exploit f's relatively small change along $\mathbf{z}$. In particular, since changes along $\mathbf{z}$ are, on average, smaller than changes along $\mathbf{y}$, we could employ a cheap model of f along $\mathbf{z}$. For example, we could replace f with an approximation $\tilde{f}$ from (4.81):

$$g(\mathbf{y}) = \left\{ \begin{array}{ll} \underset{\mathbf{z}}{\text{minimum}} & \tilde{f}(W_1 \mathbf{y} + W_2 \mathbf{z}), \\ \text{subject to} & -1 - W_1 \mathbf{y} \leq W_2 \mathbf{z} \leq 1 - W_1 \mathbf{y}. \end{array} \right. \qquad (4.84)$$

How might we choose $\tilde{f}$? One idea is to limit the number of iterations in a standard method—e.g., Newton's method—along $\mathbf{z}$ so that each iteration along $\mathbf{y}$ uses a fixed amount of work.

Another idea is to use the set of function evaluations $\{f(\mathbf{x}_j)\}$ computed while discovering the active subspace in Algorithm 3.1 to build a cheap global approximation. Suppose we have enough pairs $(\mathbf{x}_j, f_j)$ to fit the coefficients c, $\mathbf{b}$, and A of a quadratic polynomial approximation,

$$f(\mathbf{x}) \approx \tilde{f}(\mathbf{x}) = c + \mathbf{x}^T \mathbf{b} + \frac{1}{2} \mathbf{x}^T A \mathbf{x}. \qquad (4.85)$$

Using this $\tilde{f}$ makes (4.84) a quadratic program in $\mathbf{z}$.

It takes at least $\binom{m+2}{2}$ pairs $(\mathbf{x}_j, f_j)$—with the $\mathbf{x}_j$'s poised for quadratic approximation—to fit a quadratic polynomial in all m variables. If the number M of pairs is smaller than $\binom{m+2}{2}$, then we may fit a quadratic polynomial of active variables and the first few inactive variables. The additional degrees of freedom enable the optimization in (4.84). Suppose there is an integer $n' > n$ such that $\binom{n'+2}{2} \leq M$. Let W_1' be the first n' columns of the eigenvectors W from (3.5), and define the variables $\mathbf{y}' = (W_1')^T \mathbf{x}$, which contain the n active variables and the first $n' - n$ inactive variables. Use the pairs $\{(W_1')^T \mathbf{x}_j, f_j\}$ to fit the coefficients c_r, $\mathbf{b}_r$, and A_r of the quadratic polynomial,

$$g(\mathbf{y}') = c_r + (\mathbf{y}')^T \mathbf{b}_r + \frac{1}{2} (\mathbf{y}')^T A_r \mathbf{y}'. \qquad (4.86)$$

This construction is the same as that proposed in Algorithm 4.1. With these coefficients, we can construct $\tilde{f}$ as

$$\tilde{f}(\mathbf{x}) = c_r + \mathbf{x}^T \underbrace{W_1' \mathbf{b}_r}_{\mathbf{b}} + \frac{1}{2} \mathbf{x}^T \underbrace{W_1' A_r (W_1')^T}_{A} \mathbf{x}. \qquad (4.87)$$

Just like $\tilde{f}$ from (4.85), the optimization (4.84) with $\tilde{f}$ from (4.87) becomes a quadratic program in $\mathbf{z}$, but with different coefficients.

In practice, we imagine that the active subspace can help us to find a good starting point for a more general optimization method for f. We are studying the trade-offs between (i) using a standard optimization approach and (ii) discovering and exploiting the active subspace for optimization. Could the active subspace accelerate the optimization enough to justify the cost of discovering it? We can evaluate the independent samples of the gradient from Algorithm 3.1 in parallel, as opposed to sequential evaluations in a standard optimization. But the sequence of evaluations is constructed to minimize f. Many open and interesting questions remain.

4.5 ▪ Inversion

Statistical inversion seeks to characterize the uncertainty in the simulation's inputs given noisy measurements of the outputs. This is related to calibration, where the scientist seeks input values that make the simulation's output match given measurements. However, instead of finding a single value for the inputs—which can be written as a deterministic optimization—the statistical inverse problem uses probabilistic descriptions. Several excellent references describe the mathematical setup—particularly the Bayesian framework—and computational methods for the statistical inverse problem [118, 73].

Let $\mathbf{d} \in \mathbb{R}^d$ be a random vector representing the noisy measurements. We model the randomness in $\mathbf{d}$ with an additive noise model,

$$\mathbf{d} = \mathbf{m}(\mathbf{x}) + \varepsilon, \qquad (4.88)$$

where $\mathbf{x}$ denotes the simulation inputs, $\mathbf{m} : \mathbb{R}^m \to \mathbb{R}^d$ represents the simulation,[4] and ε represents the noise in the measurements. For concreteness, we assume that the noise is Gaussian with mean zero and full-rank covariance matrix $\Gamma_\mathbf{d}$, $\varepsilon \sim N(0, \Gamma_\mathbf{d})$.

In the Bayesian framework, inputs and measurements are random variables. The prior density $\rho_{\mathrm{pr}}(\mathbf{x})$ quantifies the known information on the inputs—such as their ranges and probabilities of taking particular values. We assume that the inputs have been normalized so that $\rho_{\mathrm{pr}}(\mathbf{x})$ is a Gaussian density with zero mean and covariance $\sigma^2 I$,

$$\rho_{\mathrm{pr}}(\mathbf{x}) = \frac{1}{c_{\mathrm{pr}}} \exp\left(\frac{-1}{2\sigma^2} \mathbf{x}^T \mathbf{x}\right), \qquad c_{\mathrm{pr}} = \int \exp\left(\frac{-1}{2\sigma^2} \mathbf{x}^T \mathbf{x}\right) d\mathbf{x} = \left(2\pi\sigma^2\right)^{\frac{m}{2}}. \qquad (4.89)$$

In practice, this may involve shifting the inputs by the prior mean and linearly transforming the inputs with the Cholesky factor of the prior covariance matrix.

The likelihood function of the measurements $\mathbf{d}$ given the inputs $\mathbf{x}$ is derived from the additive noise model. Rearranging (4.88), we get

$$\varepsilon = \mathbf{d} - \mathbf{m}(\mathbf{x}) \sim N(0, \Gamma_\mathbf{d}). \qquad (4.90)$$

---

[4] In this notation, the simulation may produce several quantities that can be compared to the given measurements.

Then the likelihood $\rho_{\text{lik}}(\mathbf{d}; \mathbf{x})$ is

$$\rho_{\text{lik}}(\mathbf{d}; \mathbf{x}) = \exp\left(\frac{-1}{2}(\mathbf{d} - \mathbf{m}(\mathbf{x}))^T \Gamma_{\mathbf{d}}^{-1}(\mathbf{d} - \mathbf{m}(\mathbf{x}))\right). \tag{4.91}$$

The negative of the expression inside the exponential function is called the *misfit*, which we denote by $f(\mathbf{x})$:

$$f(\mathbf{x}) = \frac{1}{2}(\mathbf{d} - \mathbf{m}(\mathbf{x}))^T \Gamma_{\mathbf{d}}^{-1}(\mathbf{d} - \mathbf{m}(\mathbf{x})). \tag{4.92}$$

Then $\rho_{\text{lik}}(\mathbf{d}; \mathbf{x}) = \exp(-f(\mathbf{x}))$, where f's dependence on $\mathbf{d}$ is implied.

The posterior density is a conditional density on the inputs given measurements corresponding to the outputs; we denote the posterior as $\rho_{\text{post}}(\mathbf{x}|\mathbf{d})$. The measurements introduce additional information that updates knowledge of the inputs. The relationship between the posterior and the prior is given by Bayes' theorem,

$$\rho_{\text{post}}(\mathbf{x}|\mathbf{d}) = \frac{1}{c_{\text{post}}}\rho_{\text{lik}}(\mathbf{d}; \mathbf{x})\rho_{\text{pr}}(\mathbf{x}) = \frac{1}{c_{\text{post}}}\exp(-f(\mathbf{x}))\rho_{\text{pr}}(\mathbf{x}), \tag{4.93}$$

where

$$c_{\text{post}} = \int \exp(-f(\mathbf{x}))\rho_{\text{pr}}(\mathbf{x})\,d\mathbf{x}. \tag{4.94}$$

The high-dimensional integral in the denominator of (4.93) is generally difficult to compute. However, without estimating the integral, we know the posterior up to a normalizing constant, which is sufficient for using Markov chain Monte Carlo (MCMC) methods to draw samples from the posterior [14, 73]. MCMC methods construct a Markov chain whose stationary distribution is equal to the posterior distribution. A path from this chain contains correlated samples from the posterior density. These samples are used to characterize the posterior, e.g., by estimating moments or constructing kernel density estimates.

It is well known that MCMC methods struggle when the dimension m of the inputs $\mathbf{x}$ is more than a handful. In essence, the chain must explore the high-dimensional space to uncover all regions where the posterior is relatively large. This high-dimensional exploration requires many iterations of the Markov chain. Each iteration uses at least one new simulation run at a new input value. If the simulation is expensive, then converged estimates of the posterior are infeasible. Accelerating convergence of MCMC methods in high dimensions with expensive forward models is a very active area of research [62, 122, 2].

If the forward problem is a PDE, then the likelihood's Hessian relates to the differential operators of the PDE, and techniques for PDE-constrained optimization can accelerate MCMC methods [17, 88]. Alternatively, recent dimension reduction ideas distinguish a data-informed subspace of the input space from a prior-informed subspace so that Markov chain iterations can focus on exploring the subspace informed by the given measurements [42]. The subspaces come from an eigenvalue decomposition of the likelihood's Hessian with respect to the inputs $\mathbf{x}$—averaged over the posterior density. Eigenvectors associated with large eigenvalues define a subspace informed by the data; the subspace defined by the remaining eigenvectors is prior-informed.

We propose using the active subspace to pursue similar dimension reduction goals. The scalar function defining the active subspace is the misfit f from (4.92), and the density $\rho(\mathbf{x})$ is the prior density. The gradient of f is

$$\nabla_{\mathbf{x}} f(\mathbf{x}) = \Gamma_{\mathbf{d}}^{\frac{-1}{2}} \nabla_{\mathbf{x}} \mathbf{m}(\mathbf{x})^T (\mathbf{d} - \mathbf{m}(\mathbf{x})), \tag{4.95}$$

where $\nabla_{\mathbf{x}}\mathbf{m} \in \mathbb{R}^{d \times m}$ is the Jacobian of the simulation outputs with respect to the inputs. One can draw samples $\mathbf{x}$ from the prior and compute the corresponding gradient independently and in parallel. So, discovering the active subspace happens before any sequential MCMC chains begin. This is potentially an advantage over sampling $\mathbf{x}$ from the posterior to estimate the data-informed subspace as in other dimension reduction approaches [42].

To exploit the active subspace, we approximate f using the conditional expectation g from (4.15). The approximation g defines an approximate likelihood, which defines an approximate posterior. The approximate posterior is much easier to sample. To see this, first note that the prior ρ_{pr} separates into a product of a prior on the active variables $\mathbf{y}$ and the inactive variables $\mathbf{z}$,

$$
\begin{aligned}
\rho_{\mathrm{pr}}(\mathbf{x}) &= \rho_{\mathrm{pr}}(W_1\mathbf{y} + W_2\mathbf{z}) \\
&= \left(2\pi\sigma^2\right)^{\frac{-m}{2}} \exp\left(\frac{-1}{2\sigma^2}(W_1\mathbf{y} + W_2\mathbf{z})^T(W_1\mathbf{y} + W_2\mathbf{z})\right) \\
&= \left(2\pi\sigma^2\right)^{\frac{-n}{2}} \exp\left(\frac{-1}{2\sigma^2}\mathbf{y}^T\mathbf{y}\right)\left(2\pi\sigma^2\right)^{\frac{-(m-n)}{2}} \exp\left(\frac{-1}{2\sigma^2}\mathbf{z}^T\mathbf{z}\right) \\
&= \rho_{\mathrm{pr},Y}(\mathbf{y})\,\rho_{\mathrm{pr},Z}(\mathbf{z}).
\end{aligned}
\tag{4.96}
$$

In other words, independence between the active and inactive variables follows from the choice of the prior. Then the approximate posterior is

$$
\begin{aligned}
\rho_{\mathrm{post}}(\mathbf{x}|\mathbf{d}) &\approx \tilde{\rho}_{\mathrm{post}}(\mathbf{x}|\mathbf{d}) \\
&= \frac{1}{\tilde{c}_{\mathrm{post}}} \exp(-g(W_1^T\mathbf{x}))\,\rho_{\mathrm{pr}}(\mathbf{x}) \\
&= \frac{1}{\tilde{c}_{\mathrm{post}}} \exp(-g(\mathbf{y}))\,\rho_{\mathrm{pr},Y}(\mathbf{y})\,\rho_{\mathrm{pr},Z}(\mathbf{z}) \\
&= \rho_{\mathrm{post},Y}(\mathbf{y}|\mathbf{d})\,\rho_{\mathrm{pr},Z}(\mathbf{z}),
\end{aligned}
\tag{4.97}
$$

where

$$
\tilde{c}_{\mathrm{post}} = \int \exp(-g(W_1^T\mathbf{x}))\,\rho_{\mathrm{pr}}(\mathbf{x})\,d\mathbf{x}.
\tag{4.98}
$$

The approximate posterior is a product of the posterior on the data-informed active variables $\mathbf{y}$ and the prior on the inactive variables $\mathbf{z}$. This implies that $\mathbf{z}$ is independent of $\mathbf{d}$. Knowledge of $\mathbf{d}$ has no effect on the knowledge of $\mathbf{z}$. In other words, $\mathbf{z}$ are unidentifiable.

An MCMC method can sample from the approximate posterior $\tilde{\rho}_{\mathrm{post}}$ in just the n dimensions of the active variables. The $\mathbf{z}$ are drawn independently of $\mathbf{y}$. We expect the chain to mix better, since perturbations in $\mathbf{y}$ change the misfit and, hence, the likelihood. Thus we expect better convergence behavior for MCMC methods with the approximate posterior. Below is an adaptation of the Metropolis–Hastings method from Kaipio and Somersalo [73, section 3.6.2] that exploits the active subspace.

ALGORITHM 4.2. Markov chain Monte Carlo with the active subspace.

1. Pick initial values $\mathbf{y}_1$, $\mathbf{z}_1$, and $\mathbf{x}_1 = W_1\mathbf{y}_1 + W_2\mathbf{z}_1$. Set $k = 1$.

2. Draw $\mathbf{y}' \in \mathbb{R}^n$ from a symmetric proposal distribution.

3. Compute the acceptance ratio

$$\gamma(\mathbf{y}_k, \mathbf{y}') = \text{minimum}\left(1, \frac{\exp(-g(\mathbf{y}'))\rho_{\text{pr},Y}(\mathbf{y}')}{\exp(-g(\mathbf{y}_k))\rho_{\text{pr},Y}(\mathbf{y}_k)}\right). \tag{4.99}$$

4. Draw t uniformly from $[0, 1]$.

5. If $\gamma(\mathbf{y}_k, \mathbf{y}') \geq t$, set $\mathbf{y}_{k+1} = \mathbf{y}'$. Otherwise, set $\mathbf{y}_{k+1} = \mathbf{y}_k$.

6. Draw $\mathbf{z}_{k+1}$ from $\rho_{\text{pr},Z}$, and compute $\mathbf{x}_{k+1} = \mathbf{W}_1 \mathbf{y}_{k+1} + \mathbf{W}_2 \mathbf{z}_{k+1}$.

7. Increment k and return to step 2.

The Markov chain in Algorithm 4.2 moves only in the n-dimensional space of the active variables, and the inactive variables are sampled independently. We are currently working to implement these ideas in various test problems. Note that the acceptance ratio in (4.99) is written for clarity—not numerical implementation.

How far is the approximate posterior from the true posterior, and is this error controlled by the eigenvalues of C? There are several metrics for distances between probability density functions. One of the most widely used is the Hellinger distance [78]. Using the results from section 4.2, we can bound the Hellinger distance between the true and approximate posteriors. We hope to extend this result to comparable measures, such as the Kullback–Leibler divergence.

Theorem 4.9. *The Hellinger distance $H(\rho_{\text{post}}, \tilde{\rho}_{\text{post}})$ between ρ_{post} and $\tilde{\rho}_{\text{post}}$ from (4.97) is bounded by a constant times the sum of the eigenvalues associated with the inactive subspace,*

$$H(\rho_{\text{post}}, \tilde{\rho}_{\text{post}}) \leq \frac{C_1^2}{4(c_{\text{post}}\tilde{c}_{\text{post}})^{\frac{1}{2}}}(\lambda_{n+1} + \cdots + \lambda_m), \tag{4.100}$$

where C_1 is from Theorem 4.3.

Proof. In what follows, we omit the explicit dependence on $\mathbf{x}$ to keep the notation clean.

$$H(\rho_{\text{post}}, \tilde{\rho}_{\text{post}}) = \frac{1}{2}\int \left((\rho_{\text{post}})^{\frac{1}{2}} - (\tilde{\rho}_{\text{post}})^{\frac{1}{2}}\right)^2 d\mathbf{x} \tag{4.101}$$

$$= \frac{1}{2}\int \left(\left(\frac{\exp(-f)\rho_{\text{pr}}}{c_{\text{post}}}\right)^{\frac{1}{2}} - \left(\frac{\exp(-g)\rho_{\text{pr}}}{\tilde{c}_{\text{post}}}\right)^{\frac{1}{2}}\right)^2 d\mathbf{x} \tag{4.102}$$

$$= \frac{1}{2}\int \left(\left(\frac{\exp(-f)}{c_{\text{post}}}\right)^{\frac{1}{2}} - \left(\frac{\exp(-g)}{\tilde{c}_{\text{post}}}\right)^{\frac{1}{2}}\right)^2 \rho_{\text{pr}}\, d\mathbf{x} \tag{4.103}$$

$$= \frac{1}{2(c_{\text{post}}\tilde{c}_{\text{post}})^{\frac{1}{2}}}\left[\int \left((\exp(-f))^{\frac{1}{2}} - (\exp(-g))^{\frac{1}{2}}\right)^2 \rho_{\text{pr}}\, d\mathbf{x}\right. \tag{4.104}$$

$$\left. - \left(\left(\int \exp(-f)\rho_{\text{pr}}\, d\mathbf{x}\right)^{\frac{1}{2}} - \left(\int \exp(-g)\rho_{\text{pr}}\, d\mathbf{x}\right)^{\frac{1}{2}}\right)^2\right] \tag{4.105}$$

$$\leq \frac{1}{2(c_{\text{post}}\tilde{c}_{\text{post}})^{\frac{1}{2}}} \int \left((\exp(-f))^{\frac{1}{2}} - (\exp(-g))^{\frac{1}{2}} \right)^2 \rho_{\text{pr}}\, d\mathbf{x} \tag{4.106}$$

$$= \frac{1}{2(c_{\text{post}}\tilde{c}_{\text{post}})^{\frac{1}{2}}} \int \left(\exp\left(\frac{-f}{2}\right) - \exp\left(\frac{-g}{2}\right) \right)^2 \rho_{\text{pr}}\, d\mathbf{x} \tag{4.107}$$

$$\leq \frac{1}{4(c_{\text{post}}\tilde{c}_{\text{post}})^{\frac{1}{2}}} \int (f - g)^2 \, \rho_{\text{pr}}\, d\mathbf{x} \tag{4.108}$$

$$\leq \frac{C_1^2}{4(c_{\text{post}}\tilde{c}_{\text{post}})^{\frac{1}{2}}} (\lambda_{n+1} + \cdots + \lambda_m). \tag{4.109}$$

Line (4.101) is the definition of the Hellinger distance. Line (4.102) plugs in the definitions of the posterior and approximate posterior in terms of the misfit and its approximation. Line (4.103) factors out the prior. Lines (4.104) and (4.105) are verified by inspection. Line (4.106) follows from the positivity of the omitted squared term. Line (4.108) follows from the boundedness of the integrand. The last line follows from Theorem 4.3. $\square$

The theorem implies that if $\lambda_{n+1} = \cdots = \lambda_m = 0$, then sampling from the approximate posterior with MCMC methods in n dimensions is equivalent to sampling from the true posterior. We expect that analyses similar to those in section 4.2 will extend this result beyond its current idealized form to more practical cases including perturbed eigenvectors $\hat{W}_1$ and approximate conditional averages $\hat{g}$.

Chapter 5

Active Subspaces in Action

This chapter presents three engineering applications where active subspaces are helping to answer questions about the relationship between a complex simulation's many inputs and its output.

5.1 ▪ HyShot II scramjet

The aerospace community has renewed its interest in numerical simulations of high-speed air-breathing propulsion systems for hypersonic scramjets. These systems are an economic alternative to rockets because they do not require an on-board oxidizer. The HyShot II is a particular scramjet design that has undergone extensive experiments [55], which makes it a ripe case for numerical study. The diagram in Figure 5.1 shows the main physical phenomena of one half of the HyShot II geometry.

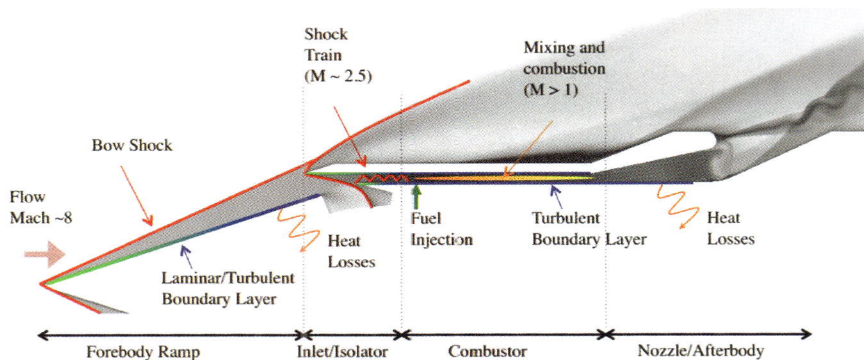

Figure 5.1. *Schematic of the main physical phenomena in the HyShot II scramjet. Reprinted with permission from Elsevier.* [70]

The supersonic flow in the combustor channel sets up a shock train that interacts with injected fuel to combust and produce thrust. However, if too much fuel is injected, the flow becomes subsonic and the thrust dissipates; this phenomenon is called *unstart*. The pressure at the end of the combustor serves as a proxy to identify unstart; we define the unstart proxy as the normalized integral of pressure over the last 10 mm of the combustor,

$$\frac{1}{\text{vol}(V)} \iiint_V P \, dx \, dy \, dz, \tag{5.1}$$

where V is the volume between $0.64 \leq x \leq 0.65$ m. This is the scalar quantity of interest we use to develop the active subspace.

The HyShot II simulation approximates the high-speed flow in the combustor with a finite volume discretization of the Reynolds-averaged Navier–Stokes equations with the $k - \omega$ turbulence model. The turbulent combustion is modeled with a flamelet/progress variable approach and coupled to the flow solver. Our previous work and its references provide more details of the simulation [29].

The simulation has seven independent input parameters that characterize the inflow boundary conditions for the combustor. Table 5.1 displays each of the seven inputs and the ranges the HyShot II may experience during operation. We motivate and characterize these parameters and their ranges in [29].

Table 5.1. *Summary of parameters and ranges used in the analysis of the HyShot II scramjet.*

Parameter	Min	Nominal	Max	Units
Stagnation pressure	16.448	17.730	19.012	MPa
Stagnation enthalpy	3.0551	3.2415	3.4280	MJ/kg
Angle of attack	2.6	3.6	4.6	deg.
Turbulence intensity	0.001	0.01	0.019	.
Turbulence length scale	0.1325	0.245	0.3575	m
Ramp transition location	0.087	0.145	0.203	m
Cowl transition location	0.030	0.050	0.070	m

The objective is to determine the effects of the seven uncertain input parameters from Table 5.1 on the scramjet's exit pressure (5.1). Let f denote the exit pressure, and let $\mathbf{x}$ denote the seven inflow parameters with ranges normalized to the interval $[-1, 1]$. We estimate the range of exit pressures caused by the range of inputs. Mathematically, this requires two global optimizations (minimize and maximize) to reveal the exit pressure range. These optimizations are challenging for several reasons:

1. The exit pressure is a functional of the pressure field computed from the nonlinear multiphysics simulation, so we have no prior knowledge of exploitable structure like linearity or convexity.

2. Each evaluation of the objective requires an expensive simulation; each run takes approximately two hours on available resources.

3. We cannot evaluate gradients or Hessians of the objective with respect to the inputs.

4. Function evaluations contain nonnegligible numerical noise due to (i) the fixed-point iteration that solves the compressible flow model and (ii) the fixed mesh constructed to capture shocks at the nominal value for the inputs; therefore, finite difference gradients cannot be trusted.

5. Given the above considerations, a seven-dimensional parameter space is extremely large for the global optimization.

These conditions are common in design problems with expensive computer simulations, and a practical approach is to optimize with the aid of response surfaces [11, 125, 72]. However, the trouble with response surfaces in our case is knowing where in the seven-dimensional space of inputs to evaluate the objective and construct the initial surrogate. Most methods use an iterative construction that evaluates the objective at a new set of points to improve accuracy near the predicted optimum. Unfortunately, this is not feasible for the scramjet, since each simulation must be carefully monitored for convergence; automating the selection of new simulation runs is impractical.

For HyShot II, we do not have access to the gradient of the quantity of interest with respect to the inflow parameters, and we do not trust finite difference approximations of the gradient due to the numerical noise. However, by performing only a handful of simulations, we can check for a one-dimensional active subspace with Algorithm 1.3 from Chapter 1. This check is so cheap that it should be the first diagnostic in any simulation study.

We apply Algorithm 1.3 to the exit pressure from the HyShot II as a function of the $m = 7$ input parameters with the ranges shown in Table 5.1. We repeat the study for two values—0.30 and 0.35—of the fuel equivalence ratio ϕ, which quantifies how much fuel is injected into the combustor. For each ϕ we sample the inputs and run the model $M = 14$ times. Figure 5.2 displays the sufficient summary plots. The horizontal axis of Figure 5.2 is the active variable $\hat{\mathbf{w}}^T \mathbf{x}$. A dominant monotonic trend exists in the exit pressure as a function of the derived active variable for both $\phi = 0.30$ and $\phi = 0.35$. We can exploit this trend to estimate the range of the exit pressure over all values of the seven input parameters.

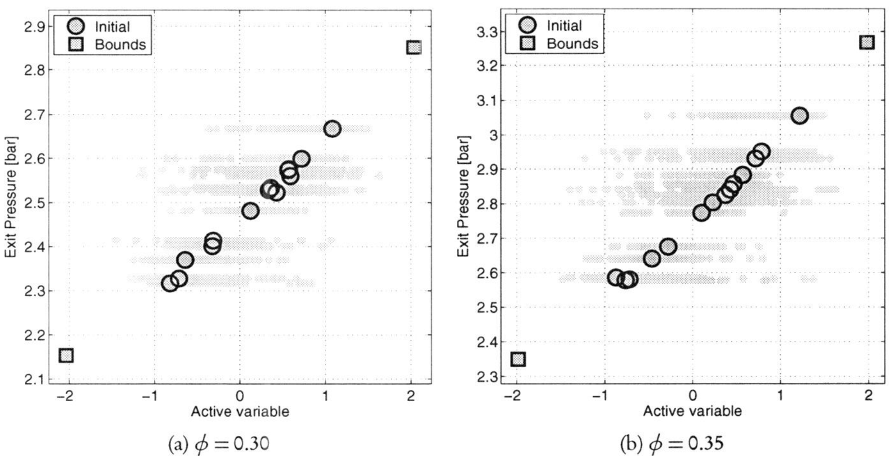

(a) $\phi = 0.30$ (b) $\phi = 0.35$

Figure 5.2. *The circles show exit pressures at $\phi = 0.30$ (a) and 0.35 (b) computed from a set of 14 HyShot II simulations and plotted against the active variable $\mathbf{w}^T \hat{\mathbf{x}}$. A clear monotonic trend is present in both cases, and we exploit it to find the range of exit pressures. The overlapping light gray circles distributed horizontally correspond to exit pressures plotted against 500 bootstrap replications of the active variable. The horizontal variability is a result of the relatively low oversampling (2x) when fitting the linear model. The gray squares show the exit pressures at the boundaries of the domain where the perceived trend suggests that we find the upper and lower bounds of the exit pressure. The monotonic trend is validated by these two additional simulations for each value of ϕ.*

The values in the vector $\hat{\mathbf{w}}$ can be used to measure the global sensitivity of the exit pressure to each of the seven parameters. These values and their corresponding input parameters are shown in Table 5.2. They suggest that four of the seven parameters contribute the most to the global variability of the exit pressure. These measures of sensitivity provide insight into the parameters that dominate the physical processes in the HyShot II simulation; see [29] for a more detailed interpretation.

Table 5.2. *The components of the vector $\hat{\mathbf{w}}$ from Algorithm 1.3 that define the active subspace computed independently for each value of ϕ. Each component corresponds to one of the uncertainty sources in the HyShot II simulation from Table 5.1.*

Index	$\phi = 0.30$	$\phi = 0.35$	Parameter
1	0.6933	0.6996	Angle of attack
2	0.5033	0.4823	Turbulence intensity
3	0.0067	−0.0270	Turbulence length scale
4	0.3468	0.1997	Stagnation pressure
5	−0.3732	−0.4738	Stagnation enthalpy
6	−0.0524	−0.0602	Cowl transition location
7	0.0605	0.0957	Ramp transition location

5.1.1 ▪ Bootstrap for $\hat{\mathbf{w}}$

The elements of $\hat{\mathbf{w}}$ depend on the samples q_i used to fit the linear model in Algorithm 1.3. With finite samples, it is natural to ask if the gradient of the linear model has been sufficiently resolved to uncover the true one-dimensional active subspace. However, with a limited budget of function evaluations, checking the convergence of $\hat{\mathbf{w}}$ is not feasible. We use a bootstrap technique [49] to estimate the variability in the computed components of $\hat{\mathbf{w}}$. The bootstrap uses 10^4 replicates. The bootstrap histograms along with stem plots of the values of $\hat{\mathbf{w}}$ from Table 5.2 are shown for $\phi = 0.30$ in Figure 5.3 and for $\phi = 0.35$ in Figure 5.4. The sharp peak of the histograms around the stems suggests confidence in the computed directions. The large ranges in the histograms are due to the low oversampling factor ($2\times$) used to fit the linear model.

To assess the variability in $\hat{\mathbf{w}}$ provided by the bootstrap, we choose a set of 500 bootstrap replicates of $\hat{\mathbf{w}}$ and plot the exit pressures against the corresponding active variables $\hat{\mathbf{w}}^T \mathbf{x}_i$. The result is a horizontal scatter of gray circles around each point in Figure 5.2. The scatter provides a visual indication of the variability in the vectors $\hat{\mathbf{w}}$. The spread relative to the range of the active variable is due to the low oversampling factor ($2\times$) used to fit the linear model.

5.1.2 ▪ Approximating the range of exit pressures

The strong monotonic trend in the exit pressure as a function of the active variable $\hat{\mathbf{w}}^T \mathbf{x}$ suggests the following simple heuristic to find the maximum and minimum exit pressures. Define

$$\mathbf{x}_{\max} = \operatorname*{argmax}_{-1 \leq \mathbf{x} \leq 1} \hat{\mathbf{w}}^T \mathbf{x}, \qquad f_{\max} = f\left(\frac{1}{2}\left[(\mathbf{x}^u - \mathbf{x}^l) \cdot \mathbf{x}_{\max} + (\mathbf{x}^u + \mathbf{x}^l) \right] \right), \qquad (5.2)$$

where $\mathbf{x}^l$ and $\mathbf{x}^u$ are the upper and lower bounds of the inputs from Table 5.1; the dot ($\cdot$) means componentwise multiplication. The scale and shift of $\mathbf{x}_{\max}$ returns it to the original

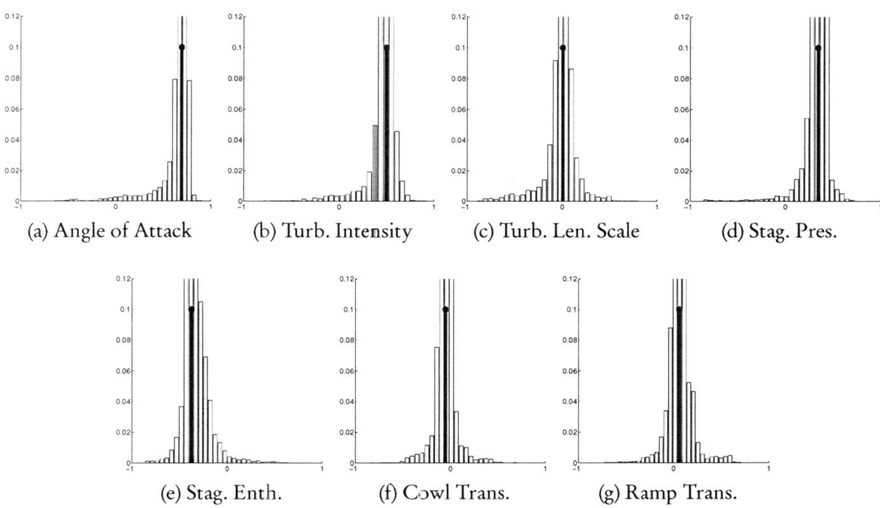

(a) Angle of Attack   (b) Turb. Intensity   (c) Turb. Len. Scale   (d) Stag. Pres.

(e) Stag. Enth.   (f) Cowl Trans.   (g) Ramp Trans.

Figure 5.3. *Bootstrap histograms of the components of the active subspace vector* $\hat{\mathbf{w}}$ *for the case where* $\phi = 0.30$. *The black stems are the computed components of* $\hat{\mathbf{w}}$ *from Table* 5.2. *The caption of each subfigure names the specific inflow parameter. The sharp peaks around each of the stems provide confidence that the computed* $\hat{\mathbf{w}}$ *are stable.*

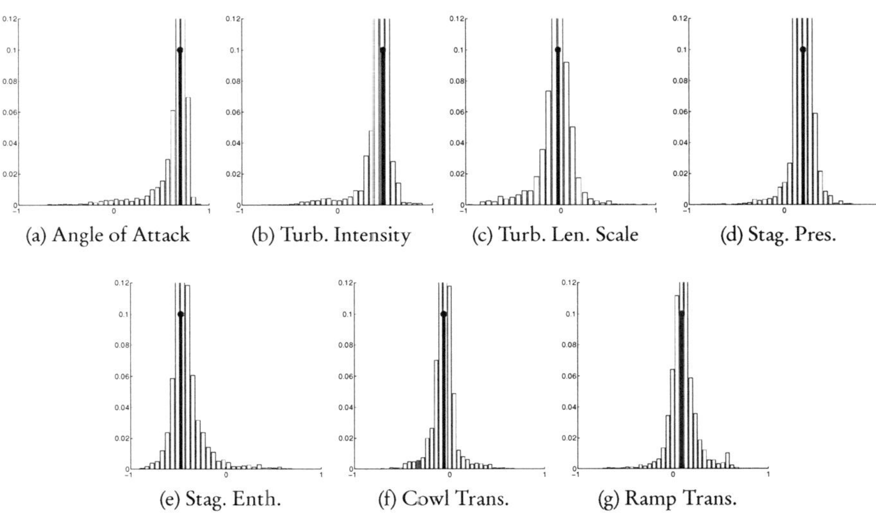

(a) Angle of Attack   (b) Turb. Intensity   (c) Turb. Len. Scale   (d) Stag. Pres.

(e) Stag. Enth.   (f) Cowl Trans.   (g) Ramp Trans.

Figure 5.4. *Bootstrap histograms of the components of the active subspace vector* $\hat{\mathbf{w}}$ *for the case where* $\phi = 0.35$. *The black stems are the computed components of* $\hat{\mathbf{w}}$ *from Table* 5.2. *The caption of each subfigure names the specific inflow parameter. The sharp peaks around each of the stems provide confidence that the computed* $\hat{\mathbf{w}}$ *are stable.*

parameter range. Similarly, define

$$\mathbf{x}_{\min} = \underset{-1 \leq \mathbf{x} \leq 1}{\operatorname{argmin}} \, \hat{\mathbf{w}}^T \mathbf{x}, \qquad f_{\min} = f\left(\frac{1}{2}\left[(\mathbf{x}^u - \mathbf{x}^l) \cdot \mathbf{x}_{\min} + (\mathbf{x}^u + \mathbf{x}^l)\right]\right). \tag{5.3}$$

The interval $[f_{\min}, f_{\max}]$ estimates the range of exit pressures from the HyShot II model. Computing these requires running the model four more times—twice for each ϕ—which is much cheaper than adaptively constructing a response surface. This is successful only because of the structure revealed in the active subspace. The maximum and minimum exit pressures are shown in Figure 5.2 as gray squares along with the runs used to determine the active subspace—all plotted against the active variable. At worst, these runs bound the initial set of 14 runs. At best, they provide estimates of the range of exit pressures over all values of the input parameters. Checking the necessary conditions for stationarity at these points is not feasible.

5.1.3 ▪ Constraining the exit pressure

To further demonstrate the utility of the active subspace, consider the following exercise in safety engineering. Suppose that the scramjet operates safely when the exit pressure is below 2.8 bar, but nears unsafe operation above 2.8 bar. With the one-dimensional active subspace, we can quickly characterize the parameter regime that produces exit pressures below the threshold of 2.8 bar.

The first step is to build a response surface model of the exit pressure as a function of the active variable $\hat{\mathbf{w}}^T \mathbf{x}$. We could try to construct a response surface of the seven model input variables. But with only 16 model runs (the first 14 samples plus the two runs used to compute the bounds), our modeling choices would be very limited. The apparent low deviation from a univariate trend in Figure 5.2 suggests that we can construct a useful response of just the active variable. In particular, we can use the 16 model evaluations for each value of ϕ and construct a univariate quadratic regression surface that models exit pressure as a function of the active variable. This construction is equivalent to a single-index regression model for the exit pressure with a quadratic polynomial link function [83].

The coefficient of determination (i.e., R^2) is 0.997 for $\phi = 0.30$ and 0.998 for $\phi = 0.35$, which gives strong confidence in the quadratic model. We use the regression model to find the largest value of the active variable such that the regression surface's upper 99% confidence limit is less than the threshold of 2.8 bar. Figure 5.5 shows the regression surface and its upper 99% confidence limit for both values of ϕ. The shaded region identifies the values of the active variable that produce exit pressures at or below the pressure threshold.

Let $y_{\max}$ be the value of the active variable where the regression's upper confidence limit crosses the pressure threshold. Then the safe region of the normalized input parameters is the set $\mathscr{S}$ defined as

$$\mathscr{S} = \{\mathbf{x} : \hat{\mathbf{w}}^T \mathbf{x} \leq y_{\max}, \, -1 \leq \mathbf{x} \leq 1\}. \tag{5.4}$$

One can easily shift and scale this region to the space of the model's true input variables. The linear inequality constraint implies that the inputs are related with respect to the exit pressure. In other words, the complete range of safe angles of attack depends on the other input variables. The presence of the active subspace and the quality of the regression surface enable us to characterize these relationships.

The safe set $\mathscr{S}$ defined in (5.4) is like a seven-dimensional box with the top chopped off by the hyperplane $\mathbf{x}^T \hat{\mathbf{w}} \leq y_{\max}$. We can identify a new set of independent ranges for

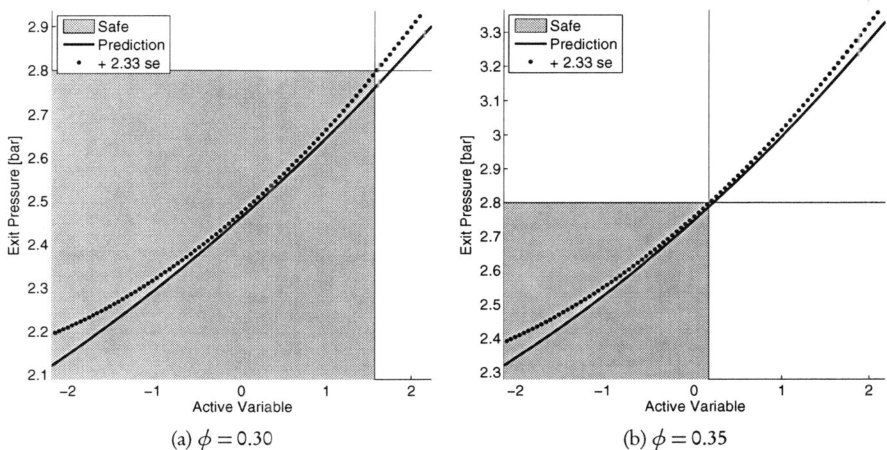

(a) $\phi = 0.30$ (b) $\phi = 0.35$

Figure 5.5. *A quadratic regression surface models the relationship between the active variable and the exit pressure. The solid line shows the mean prediction of the regression surface. The dotted line shows the upper 99% confidence limit for the prediction. We find the value of the active variable $y_{\max}$ where the upper confidence limit crosses the safety threshold of 2.8 bar. All values of the active variable less than $y_{\max}$ produce pressures within the safety limit. The set of safe input variables is shown in (5.4).*

the input variables such that all inputs within those ranges produce exit pressures below the 2.8 bar safety threshold; this is like finding the largest seven-dimensional box that fits inside the set $\mathscr{S}$. More precisely, we solve the optimization problem

$$
\begin{aligned}
\underset{\mathbf{x}}{\text{maximize}} \quad & \prod_{i=1}^{m} |x_i - x_{i,\min}|, \\
\text{subject to} \quad & \mathbf{x} \in \mathscr{S},
\end{aligned}
\tag{5.5}
$$

where $x_{i,\min}$ are the components of the minimizer from (5.3). The maximizing components define the corner of the largest hyperrectangle opposite the corner $\mathbf{x}_{\min}$.

We can interpret such analysis as backward uncertainty propagation that characterizes safe inputs given a characterization of a safe output under the constraint that the inputs be independent. For the present study, these spaces differ between the two fuel flow rates; see Figure 5.5. For $\phi = 0.30$, only the angle of attack and turbulent intensity are affected by the safety constraint on the pressure; the angle of attack must be less than 3.99 degrees, and turbulent intensity must be less than 0.018. For $\phi = 0.35$, the same parameters are constrained—angle of attack less than 3.31 degrees and turbulent intensity less than 0.011—and a stricter minimum on stagnation enthalpy (greater than 2.23 MJ/kg) appears.

Finally, we use the quadratic response surface of the active variable to estimate a cumulative distribution function of the exit pressure for both fuel injection rates. The procedure we follow is to sample uniformly from the scramjet's seven-dimensional input space, and for each sample we evaluate the active variable and the quadratic response of the active variable. Figure 5.6 shows estimates of the cumulative distribution function computed from these samples with a Gaussian kernel density estimator. The vertical lines show the estimated upper and lower bounds, $f_{\max}$ and $f_{\min}$. Discovering the low-dimensional structure with the active subspace and quantifying uncertainty requires 16 simulation runs for each of two fuel equivalence ratios. This number is remarkably small considering both the complexity of the simulation model and its dependence on seven independent input

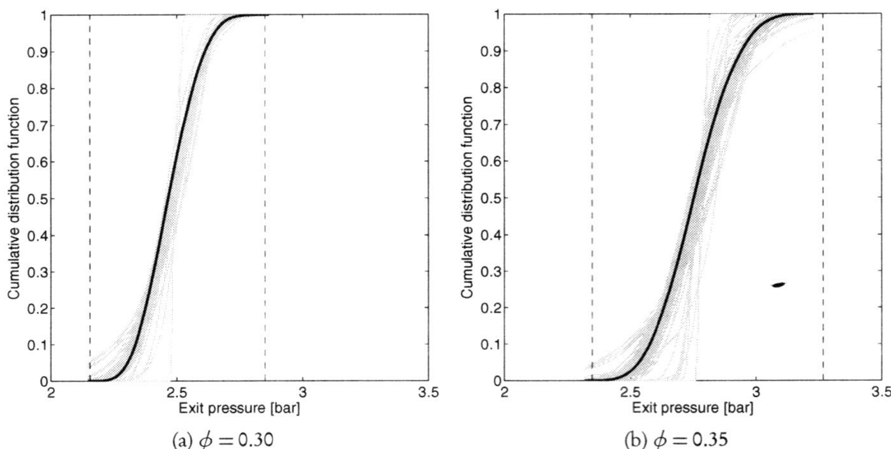

(a) $\phi = 0.30$ (b) $\phi = 0.35$

Figure 5.6. *Estimated cumulative distribution functions for exit pressure at both values of fuel equivalence ratio ϕ. These are estimated with Gaussian kernel density estimates; the samples are drawn from the quadratic response surface of the active variable. Vertical lines show the estimated bounds for each case. Each of the 500 gray lines is computed with a bootstrap replicate of the direction $\hat{\mathbf{w}}$.*

parameters. The exceptionally small computational cost is a result of the low-dimensional structure revealed by the active subspace.

5.2 ▪ Photovoltaic solar cell

Growing demand for clean energy has caused rapid growth in the photovoltaic (PV) industry. The health of the industry depends on proper assessment of a PV system's risks. One can assess these risks with a mathematical model of the device's performance, such as a single-diode lumped-parameter equivalent-circuit model. For many series-wired PV devices, the single-diode model accurately describes the device's current-voltage (I-V) characteristics at given irradiance and temperature conditions.

There are five input parameters in the single-diode model that must be calibrated for a given PV device. Once these parameters are determined—e.g., using methods from [63] or [50]—engineers can predict the device's *key performance measures*, such as the maximum power output P_{max} and energy conversion efficiency η. Calibrating the input parameters presents several challenges. For example, many calibration methods have been proposed in the PV literature for single-diode models, and it is not always clear which is most appropriate [64]. Further, errors in calibration cause errors in the device's predicted performance measures. The first step to quantifying such errors is to analyze the predictions' sensitivity to changes in the input parameters.

We use active subspaces to study how variability in the five input parameters affects variability in the single-diode model's predictions of the device's maximum power P_{max}. The single-diode model defines a relationship between current I and voltage V,

$$I = I_{\text{L}} - I_{\text{S}}\left(\exp\left(\frac{V + IR_{\text{S}}}{N_{\text{S}} n V_{\text{th}}}\right) - 1\right) - \frac{V + IR_{\text{S}}}{R_{\text{P}}}, \tag{5.6}$$

where N_{S} is the number of cells connected in series, and the thermal voltage V_{th} is a known value that depends on temperature. The photocurrent I_L satisfies an auxiliary relation,

$$I_L = E I_{SC} + I_S \left(\exp\left(\frac{E I_{SC} R_S}{N_S n V_{th}} \right) - 1 \right) + \frac{E I_{SC} R_S}{R_P}. \tag{5.7}$$

The relations (5.6) and (5.7) are valid for the standard reporting conditions, which include fixed temperature $T = 25°\,C$ and irradiance $E = 1000\,\text{W/m}^2$. The five device-specific parameters are the short-circuit current I_{SC}, the reverse saturation current I_S, the ideality factor n, the series resistance R_S, and the parallel resistance R_P. Ranges of the parameters are shown in Table 5.3; these values are typical for a $2\,\text{cm}^2$ silicon PV cell.

Table 5.3. *Ranges of the five device-specific parameters for the single-diode model. These ranges contain values that are typical of a $2\,cm^2$ silicon PV cell.*

Description	Symbol	Lower bound	Upper bound	Units
Short-circuit current	I_{SC}	0.05989	0.23958	amps
Diode reverse saturation current	I_S	2.2e-11	2.2e-7	amps
Ideality factor	n	1	2	unitless
Series resistance	R_S	0.16625	0.66500	ohms
Parallel (shunt) resistance	R_P	93.75	375.00	ohms

The maximum power of a PV device is

$$P_{max} = \underset{I,V}{\text{maximum}}\; IV, \tag{5.8}$$

where I and V are constrained by the single-diode model (5.6)–(5.7). Changes in the input parameters affect the nonlinear constraint function, which affects P_{max}. A MATLAB code that solves the constrained optimization (5.8) given values for the five parameters in Table 5.3.

Let f be the maximum power of the PV device from (5.8), and let $\mathbf{x}$ be the device-specific parameters from Table 5.3 normalized so that each component of $\mathbf{x}$ takes values in $[-1, 1]$. We set the second component of $\mathbf{x}$ to be $\log(I_{SC})$ normalized to $[-1, 1]$. The script that solves (5.8) does not provide the gradient of P_{max} with respect to the parameters. But the code is sufficiently fast and the output is sufficiently smooth for accurate finite difference approximations. We approximate the gradient by a first-order finite difference with stepsize 10^{-6}. With $\mathbf{x}$, f, and $\nabla_{\mathbf{x}} f$ defined, we now attempt to discover an active subspace.

Using Algorithm 3.3 with finite differences, we choose the multiplier $\alpha = 10$ and $k = m + 1 = 6$, which is sufficient to estimate all five eigenvalues. We draw $M = 97$ points $\mathbf{x}_i$ uniformly from the hypercube $[-1, 1]^5$, and for each point we compute P_{max} and the finite difference gradient. This takes 582 simulations.

To study variability in the estimated eigenpairs, we bootstrap with 10000 bootstrap replicates. Figure 5.7 displays the results of Algorithm 3.3. (a) shows the eigenvalue estimates and their bootstrap intervals. The order-of-magnitude gaps between the eigenvalues suggests the presence of an active subspace. (b) shows the estimated subspace error.

Figure 5.8 displays the components of the first and second estimated eigenvectors. The gray regions show the eigenvectors computed from the first 500 bootstrap replicates. The small ranges suggest that the one- and two-dimensional active subspaces are stable.

The eigenvector components connect the active subspace to the (normalized) inputs in the single-diode model. The indices on the horizontal axis in Figure 5.8 correspond

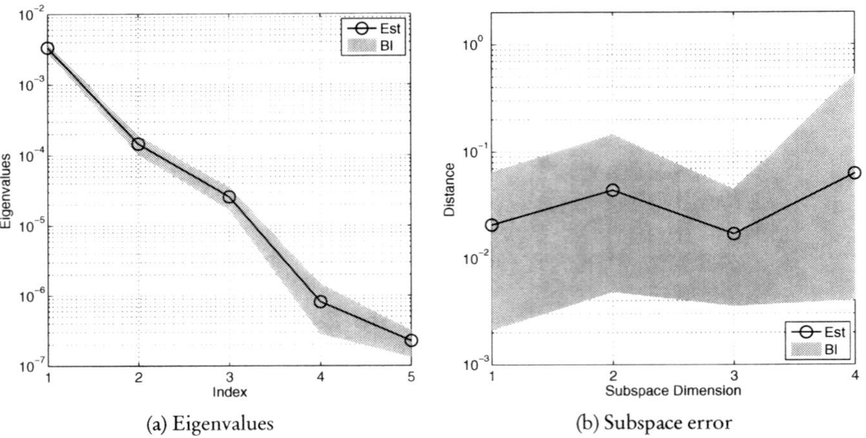

(a) Eigenvalues   (b) Subspace error

Figure 5.7. (a) *shows the eigenvalue estimates in black circles. The gray regions are the bootstrap intervals with* 10000 *bootstrap replicates. The order-of-magnitude gaps between the eigenvalues suggest confidence in the dominance of the active subspace.* (b) *shows the estimated error in subspaces of dimension one to four. The bootstrap intervals are gray.*

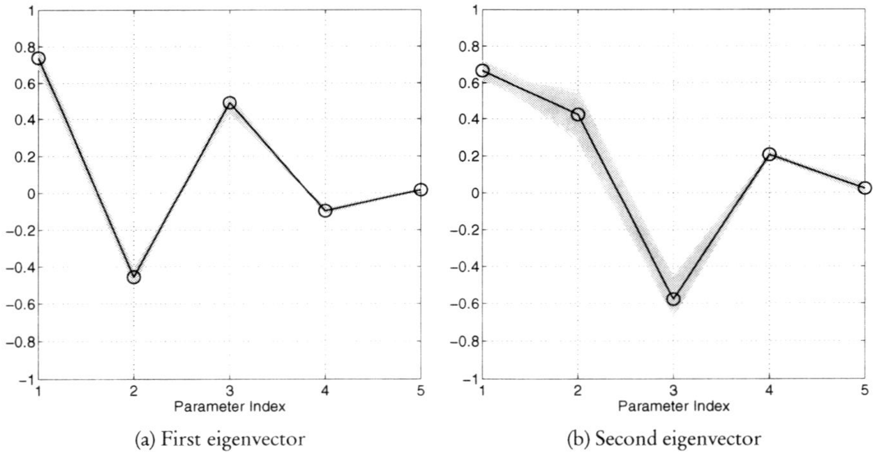

(a) First eigenvector   (b) Second eigenvector

Figure 5.8. *These figures show the components of the first* (a) *and second* (b) *eigenvectors of* $\hat{C}$ *from Algorithm* 3.3. *The gray regions are the eigenvectors from the first* 500 *bootstrap replicates of* $\hat{C}$. *The narrow ranges indicate that the estimated subspaces are stable. The labels on the horizontal axis correspond to the index of the parameters in Table* 5.3.

to the rows of Table 5.3. The magnitudes of the eigenvector components can be used as measures of relative sensitivity for each of the parameters in the model. A large absolute value of the eigenvector component implies that this variable is important in defining the direction that changes P_{max} the most, on average. The component corresponding to the parallel resistance R_{p} is close to zero, which implies that changes in R_{p} do not cause as much change to P_{max} as the other parameters. A sensitivity analysis that was limited to

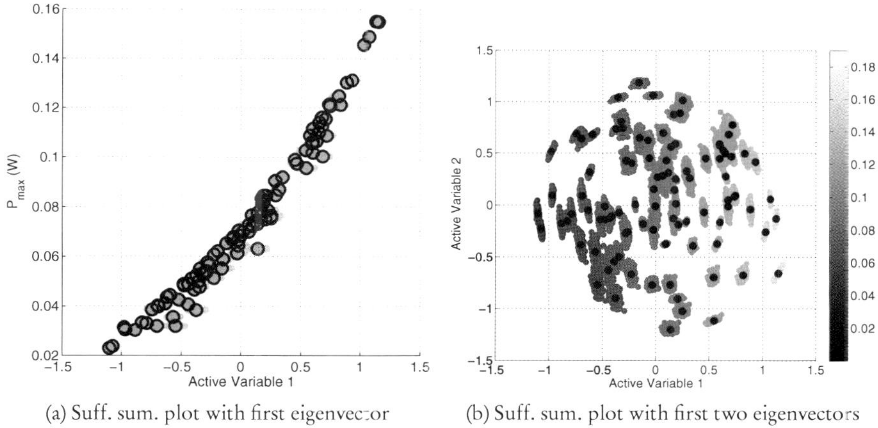

(a) Suff. sum. plot with first eigenvector   (b) Suff. sum. plot with first two eigenvectors

Figure 5.9. *Sufficient summary plots of P_{max} using the eigenvectors from $\hat{C}$ and the bootstrap replicates. The strong trend in (a) is verified by looking at the variation along the horizontal axis in (b), and the combination of these provides evidence of a dominant one-dimensional active subspace.*

model variables (e.g., variance-based decompositions that use Sobol' indices [110]) would likely conclude that four of the five coordinate dimensions were important. However, the rotation of the domain produced by the eigenvector matrix $\hat{W}$ enables a more general sensitivity analysis.

Figure 5.9(a) shows the one- and two-dimensional sufficient summary plots of the 97 P_{max} values using the first eigenvector of $\hat{C}$. The horizontally oriented clusters are superimposed summary plots using the first eigenvector from the bootstrap replicates. The tight clusters around the univariate trend confirm the dominance of the one-dimensional active subspace.

Figure 5.9(b) is a scatter plot—with grayscale corresponding to the vertical component—of the same 97 values of P_{max} and the variation from the eigenvectors' bootstrap replicates. Each shaded cluster corresponds to one value of P_{max}; the black dots use the eigenvectors of $\hat{C}$. The one-dimensional active subspace's dominance is apparent in this plot, since more variation in P_{max} is visible along the first active variable (the horizontal axis) than the second (the vertical axis).

5.3 ▪ Airfoil shape optimization

Methods in aerospace shape optimization iteratively change an aircraft's geometry to improve its performance. Realistic three-dimensional design problems parameterize the geometry with hundreds of variables to permit a wide range of geometric variation [102, 75]. But this large number of parameters makes the optimization problem more difficult, particularly since the performance objectives are not generally convex with respect to the parameters. An active subspace in a performance objective—such as the aircraft's lift or drag—can potentially accelerate the design optimization.

We use active subspaces to study the performance of two standard transonic test problems: the NACA0012 airfoil and the ONERA-M6 fixed wing. In both cases, the inputs $\mathbf{x}$ are the parameters characterizing the geometry. The $\mathbf{x}$ are constrained to ensure that all parameter values yield feasible geometries (e.g., the upper boundary is sufficiently separated from the lower boundary). The quantities of interest are lift and drag, which are derived scalars from the flow field solutions. Given a geometry, we use the Stanford Aerospace Design Laboratory SU^2 compressible flow solver [103] to approximate the velocity and pressure fields. SU^2 also contains tools for optimal shape design including adjoint solvers, free-form mesh deformation, goal-oriented adaptive mesh refinement, and a constrained optimization framework.

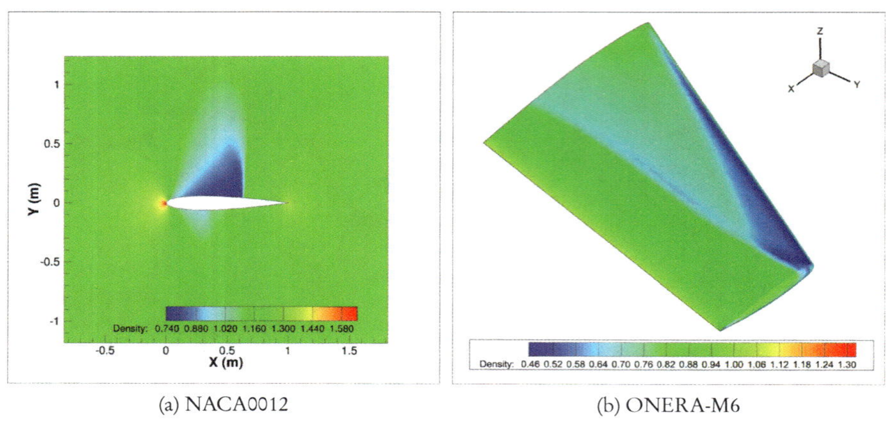

(a) NACA0012　　　　　　(b) ONERA-M6

Figure 5.10. *Density contours for the baseline geometries for the NACA0012 airfoil* (a) *and the ONERA-M6 fixed wing* (b). *The sharp gradient in the density represents the shock wave in the transonic, compressible flow field. Image* (b) *is taken from* [87] *and is reprinted with permission from AIAA.*

5.3.1 ▪ NACA0012 airfoil

The NACA0012 airfoil is a standard test problem for transonic, compressible flow in two spatial dimensions. A representative density field with the baseline geometry is shown in Figure 5.10(a). The airfoil's geometry is parameterized by 18 Hicks-Henne bump functions, where the parameters control the height of the bumps. Changes in the parameters from the nominal value of zero smoothly deform the geometry. Each parameter is constrained to the interval $[-0.01, 0.01]$, and points in the 18-dimensional hypercube produce valid geometries for flow computations. We equip the 18-dimensional hypercube—normalized to $[-1, 1]^{18}$—with a uniform density function. The quantities of interest for shape optimization are lift and drag of the airfoil. Gradients of lift and drag with respect to the shape parameters are available through the SU^2 adjoint solver.

Figure 5.11 shows the results of Algorithm 3.3 for sampling the gradient and estimating the eigenpairs that identify and define the active subspace—both for drag and lift. We choose to examine the first $k = 7$ eigenvalues, and we use a multiplier of $\alpha = 6$, which leads us to drawing $M = 122$ samples of the gradient. The bootstrap intervals—represented by the shaded regions in Figure 5.11—are computed with 1000 bootstrap replicates. The relatively large gap between the first and second eigenvalues for lift indicates that it should be well-approximated by a univariate function of the first active variable. The estimates of the subspace error suggest that the one-dimensional subspace is relatively well-approximated for lift. The eigenvalues for drag do not clearly indicate the appropriate dimension of the active subspace. The gap between the first and second eigenvalues is

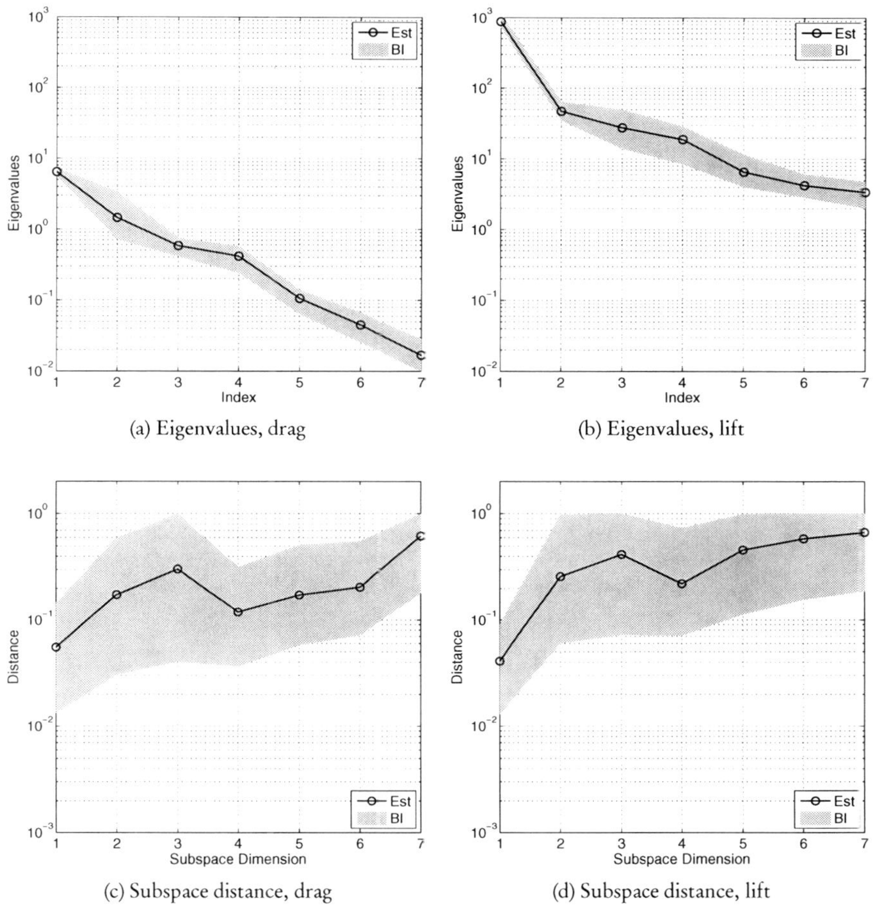

(a) Eigenvalues, drag

(b) Eigenvalues, lift

(c) Subspace distance, drag

(d) Subspace distance, lift

Figure 5.11. *Eigenvalue estimates ((a),(b)) and estimated subspace errors ((c),(d)) for both drag ((a),(c)) and lift ((b),(d)) from the NACA0012 airfoil. The relatively large gap between the first and second lift eigenvalues leads to a more accurate estimate of the one-dimensional active subspace for lift.*

close to the gap between the fourth and fifth eigenvalues, which is reflected by the relative dip in the estimates of the subspace distance. Figure 5.12 shows the one-dimensional sufficient summary plots for both drag and lift with weights from the first active subspace eigenvector. The lift can apparently be accurately modeled with a univariate function of the first active variable—even with a linear function. The spread around the mean trend is relatively small compared to the drag's spread around a mean univariate trend. This is consistent with the relatively large gap between the first and second eigenvalues for lift in Figure 5.11(b).

These low-dimensional active subspaces—one-dimensional for lift and one- or four-dimensional for drag—can potentially be exploited for shape optimization using ideas from section 4.4. This is currently work in progress.

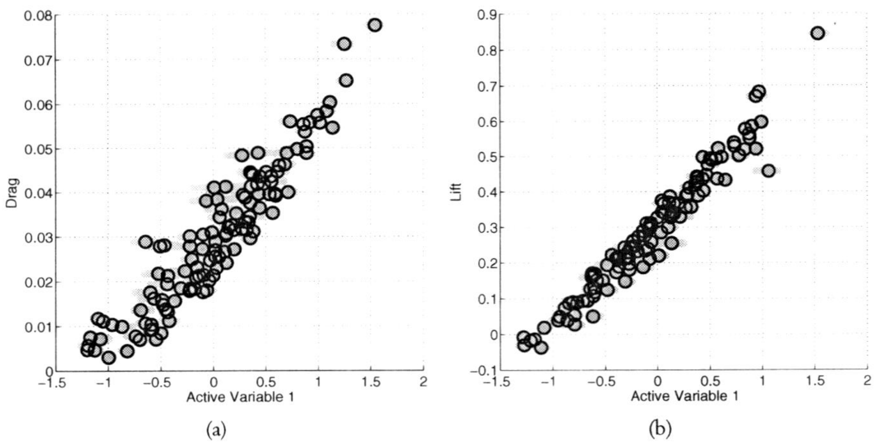

Figure 5.12. *One-dimensional sufficient summary plots with the first active variable for drag (a) and lift (b) from the NACA0012 airfoil. The relatively small spread about a univariate trend for lift corresponds to the relatively large gap between the first and second eigenvalues in Figure 5.11(b).*

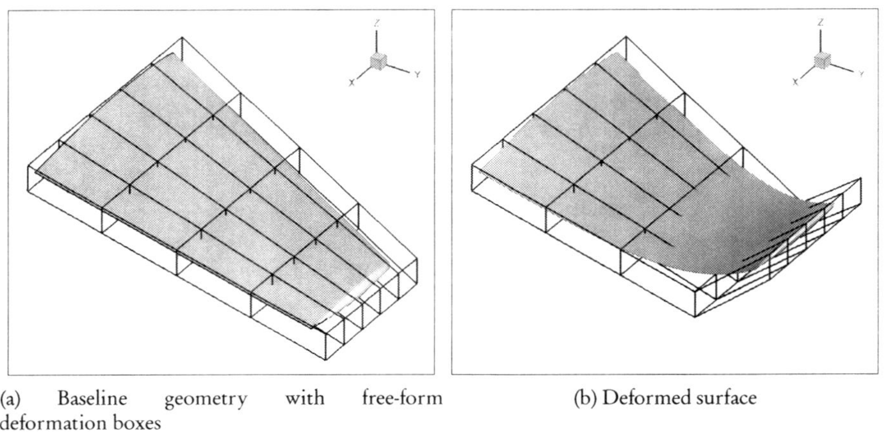

(a)   Baseline   geometry   with   free-form   (b) Deformed surface
deformation boxes

Figure 5.13. *The free-form deformation approach parameterizes the surface with continuous parameters such that changes in the parameters produce smooth changes in the surface. Reprinted from [87] with permission from AIAA.*

5.3.2 ▪ ONERA-M6 fixed wing

We turn to the ONERA-M6 test case. Details of these computations—including heuristics for exploiting the active subspace in the shape optimization—can be found in our recent work [87]. Moving from two to three spatial dimensions allows more complex discontinuities in the density field. Contours of the density field on the wing surface for the baseline geometry are shown in Figure 5.10(b). The geometry parameterization employs free-form deformation boxes. Varying a set of control points smoothly changes the surface's thickness, sweep, or twist. Figure 5.13 shows an example deformation. There

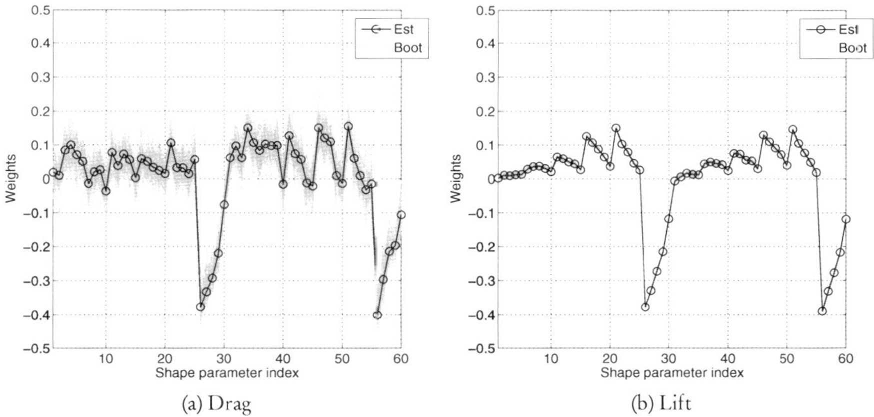

(a) Drag   (b) Lift

Figure 5.14. *Components of the vector* $\hat{\mathbf{w}}$ *produced by Algorithm* 1.3 *for drag* (a) *and lift* (b). *The gray lines show the bootstrap replicates. Note the relatively large spread around the drag's weights.*

are 60 free-form deformation parameters, and each ranges from −0.05 to 0.05; the range is chosen to ensure that all points produce valid wings. The normalized parameter space is the hypercube $[-1,1]^{60}$ equipped with a uniform probability density. The quantities of interest are the lift and drag of the wing. Unfortunately, at the time we ran the experiments the SU^2 adjoint solver struggled with the sharp trailing edge of this geometry, so we do not have access to gradients of lift and drag with respect to the free-form deformation parameters. We use Algorithm 1.3—which uses a global linear approximation of the quantity of interest—to search for a one-dimensional active subspace. We fit the linear models of lift and drag as functions of the 60 normalized free-form deformation parameters using 248 runs with randomly perturbed geometries. Figure 5.14 shows the components of the vector $\hat{\mathbf{w}}$ from Algorithm 1.3 that defines the one-dimensional active subspace for drag (Figure 5.14(a)) and lift (Figure 5.14(b)). The gray lines show bootstrap replicates. Notice the bootstrap's relatively large spread for drag. Also notice how similar the weights are for the two quantities of interest. Geometric perturbations that increase the wing's lift often also increase its drag. Figures 5.15(a)–(c) show the sufficient summary plot for drag. The plot repeats three times with a vertical line at specific values of the active variable. The figures below each sufficient summary plot show the wing's geometric changes as the active variable sweeps across its range; the color indicates deviation from the baseline geometry. This one-dimensional geometric perturbation changes lift more, on average, than other perturbations. This perturbation confirms the engineering intuition that thicker wings create more drag, and the trailing edge contributes significantly to drag. Figures 5.16(a)–(c) present the sufficient summary plot for lift, which shows that lift is very nearly a linear function of its first active variable. There is very little visible deviation from a univariate trend. This is consistent with the relatively small spread in the weight vector's bootstrap replicates in Figure 5.14(b). Again, the vertical lines indicate a specific value of the active variable, and Figures 5.16(d)–(f) show the wing's corresponding geometric perturbations. The shape of the trailing edge contributes the most to changes in the wing's lift, which matches the engineering intuition.

Active subspaces have provided significant insight into the behavior of lift and drag as functions of the parameters characterizing the shape for both the NACA0012 and the

ONERA-M6. The story is similar for both cases. The lift is relatively well-approximated by a linear function of its first active variable, while the drag's variation needs more linear combinations of the shape parameters to describe its functional relationship.

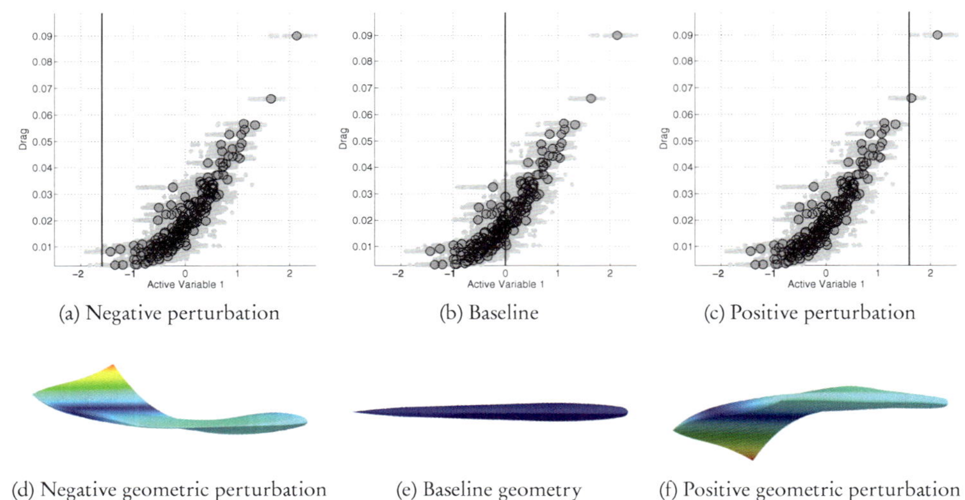

(a) Negative perturbation   (b) Baseline   (c) Positive perturbation

(d) Negative geometric perturbation   (e) Baseline geometry   (f) Positive geometric perturbation

Figure 5.15. *(a)–(c) contain the sufficient summary plots for the ONERA-M6 drag. The plots are identical except for a vertical line indicating the active variable's value in the geometric perturbation below it. (d)–(f) show the wing's geometric perturbation corresponding to a large change in the active variable. (b),(e) show the baseline geometry, and (a),(d) and (c),(f) show positive and negative perturbations, respectively, of the active variable.*

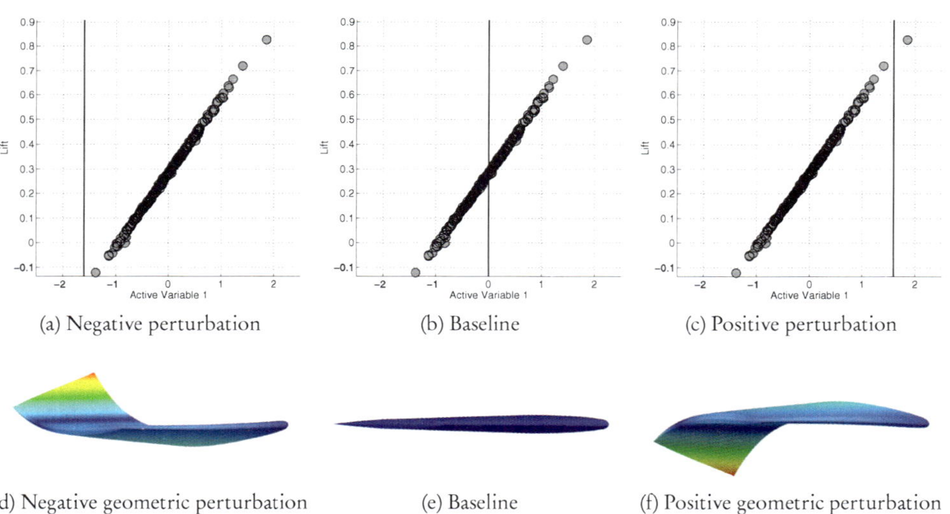

(a) Negative perturbation   (b) Baseline   (c) Positive perturbation

(d) Negative geometric perturbation   (e) Baseline   (f) Positive geometric perturbation

Figure 5.16. *This figure shows the same plots as Figure 5.15 except with lift as the quantity of interest defining the active variable.*

Chapter 6

Summary and Future Directions

The active subspace is defined by a set of directions in a multivariate function's input space. Perturbing the inputs along these directions changes the function's output more, on average, than perturbing the inputs in orthogonal directions. The active subspace is derived from the function's gradient, so it is a property of the function. We have presented and analyzed a technique based on randomly sampling the gradient to test whether a function admits an active subspace. When an active subspace is present, one can exploit it to reduce the dimension for computational studies that seek to characterize the relationship between the function's inputs and outputs, such as optimization, uncertainty quantification, and sensitivity analysis. This is especially useful for complex engineering simulations with more than a handful of inputs.

There are several paths to explore in active subspaces. The ideas are still emerging, so they remain open for innovation and improvement. Several of the concepts in Chapter 4 are ripe for extension and further analysis. The results in this book are beginning to lay the groundwork for a set of tools that computational scientists across the engineer-to-analyst spectrum will find useful. We mention a few ideas that repeatedly come up in our conversations about active subspaces.

6.1 ▪ Multiple outputs

All definitions and derivations assume that f is a scalar-valued function. If a simulation has multiple quantities of interest, then one can independently search for an active subspace in each quantity. But this approach does not translate to cases where f's output is a spatially varying field or a time series. For example, suppose $f = f(t, \mathbf{x})$, where t represents time. One approach might be to study the time dependent matrix

$$C(t) = \int \nabla_{\mathbf{x}} f(t, \mathbf{x}) \nabla_{\mathbf{x}} f(t, \mathbf{x})^T \rho \, d\mathbf{x}. \tag{6.1}$$

Another approach might be to study the constant matrix with block structure,

$$\int \int \begin{bmatrix} \left(\frac{\partial f}{\partial t}\right)^2 & \frac{\partial f}{\partial t} \nabla_{\mathbf{x}} f^T \\ \nabla_{\mathbf{x}} f \frac{\partial f}{\partial t} & \nabla_{\mathbf{x}} f \nabla_{\mathbf{x}} f^T \end{bmatrix} \rho(t, \mathbf{x}) \, dt \, d\mathbf{x}, \tag{6.2}$$

where ρ is a joint density on the combined time/parameter domain. The Schur complement of this matrix is $m \times m$, and it may produce a comparable active subspace.

6.2 ▪ Coupled systems

Most complex engineering systems contain several components that are coupled through interfaces; the output of one system becomes the input of another. A fully coupled system might be written

$$F(\mathbf{u}; \mathbf{v}) = 0, \qquad G(\mathbf{v}; \mathbf{u}) = 0, \tag{6.3}$$

where the system F takes inputs $\mathbf{v}$ and produces $\mathbf{u}$, and the system G takes inputs $\mathbf{u}$ and produces $\mathbf{v}$. The off-diagonal blocks of the full system's Jacobian contain F's partial derivatives with respect to its inputs $\mathbf{v}$ and G's partial derivatives with respect to its inputs $\mathbf{u}$. We may be able to analyze these Jacobians in the same way that we analyze the gradients with the matrix C from (3.2). We could find directions in the space of coupling terms where the system is strongly coupled and other directions where the system is weakly coupled—which could lead to more efficient solvers for coupled systems.

6.3 ▪ Anticipating active subspaces

One question that arises frequently in our conversations is, what sorts of applications admit active subspaces? We take a bottom-up approach to answering this question. In other words, we check for an active subspace using the tests mentioned in Chapters 1 and 3 in data sets from every application we can access. We have presented three relatively thorough investigations in Chapter 5, but we have tried many others. Check the website `activesubspaces.org` for links to the data sets we have tested. We hope that such an approach will eventually lead to general statements, but we are not there yet.

A related question is, suppose one finds an active subspace in a low-fidelity, easy-to-check model of an engineering system. Will an active subspace be present in a related high-fidelity model, where several samples are not feasible? For example, could analyzing a cheaper steady state model inform the analysis of a more expensive dynamic model? To mathematically answer this question, one needs a solid understanding of the relationship between the low-fidelity mathematical model and its high-fidelity counterpart. But this information is not always available, since the low-fidelity model may be derived from tenuously justified approximations. Nevertheless, the idea is appealing.

Bibliography

[1] H.S. ABDEL-KHALIK, Y. BANG, AND C. WANG, *Overview of hybrid subspace methods for uncertainty quantification, sensitivity analysis*, Annals of Nuclear Energy, 52 (2013), pp. 28–46. Nuclear Reactor Safety Simulation and Uncertainty Analysis. (Cited on p. 18)

[2] A. APTE, M. HAIRER, A.M. STUART, AND J. VOSS, *Sampling the posterior: An approach to non-Gaussian data assimilation*, Physica D: Nonlinear Phenomena, 230 (2007), pp. 50–64. Data Assimilation. (Cited on p. 66)

[3] V. BARTHELMANN, E. NOVAK, AND K. RITTER, *High dimensional polynomial interpolation on sparse grids*, Advances in Computational Mathematics, 12 (2000), pp. 273–288. (Cited on p. 16)

[4] M. BEBENDORF, *A note on the Poincaré inequality for convex domains*, Zeitschrift fur Analysis und Ihre Anwendungen, 22 (2003), pp. 751–756. (Cited on p. 50)

[5] M. BEBENDORF, *Adaptive cross approximation of multivariate functions*, Constructive Approximation, 34 (2011), pp. 149–179. (Cited on p. 16)

[6] S.H. BERGUIN, D. RANCOURT, AND D.N. MAVRIS, *A method for high-dimensional design space exploration of expensive functions with access to gradient information*, in 15th AIAA/ISSMO Multidisciplinary Analysis and Optimization Conference, Atlanta, GA, 2014, paper AIAA-2014-2174. (Cited on p. 18)

[7] S.H. BERGUIN AND D.N. MARVIS, *Dimensionality reduction using principal component analysis in aerodynamic design*, in 10th AIAA Multidisciplinary Design Optimization Conference, National Harbor, MD, 2004, paper AIAA 2014-0112. (Cited on p. 18)

[8] G. BERKOOZ, P. HOLMES, AND J.L. LUMLEY, *The proper orthogonal decomposition in the analysis of turbulent flows*, Annual Review of Fluid Mechanics, 25 (1993), pp. 539–575. (Cited on p. 37)

[9] G. BEYLKIN AND M.J. MOHLENKAMP, *Algorithms for numerical analysis in high dimensions*, SIAM Journal on Scientific Computing, 26 (2005), pp. 2133–2159. (Cited on p. 16)

[10] L. BIEGLER, G. BIROS, O. GHATTAS, M. HEINKENSCHLOSS, D. KEYES, B. MALLICK, Y. MARZOUK, L. TENORIO, B. VAN BLOEMEN WAANDERS, AND K. WILLCOX, *Large-Scale Inverse Problems and Quantification of Uncertainty*, West Sussex: Wiley, 2011. (Cited on p. 15)

[11] S.C. BILLUPS, J. LARSON, AND P. GRAF, *Derivative-free optimization of expensive functions with computational error using weighted regression*, SIAM Journal on Optimization, 23 (2013), pp. 27–53. (Cited on p. 73)

[12] S. BOYD AND L. VANDENBERGHE, *Convex Optimization*, Cambridge: Cambridge University Press, 2009. (Cited on p. 16)

[13] J. BREIDT, T. BUTLER, AND D. ESTEP, *A measure-theoretic computational method for inverse sensitivity problems I: Method and analysis*, SIAM Journal on Numerical Analysis, 49 (2011), pp. 1836–1859. (Cited on p. 15)

[14] S. BROOKS, A. GELMAN, G. JONES, AND X.-L. MENG, *Handbook of Markov Chain Monte Carlo*, Boca Raton, FL: CRC Press, 2011. (Cited on p. 66)

[15] A.E. BRYSON AND Y.-C. HO, *Applied Optimal Control: Optimization, Estimation, and Control*, Washington, DC: Hemisphere Publishing Corportation, 1975. (Cited on p. 14)

[16] B. BÜELER, A. ENGE, AND K. FUKUDA, *Exact volume computation for polytopes: A practical study*, in Polytopes - Combinatorics and Computation, G. Kalai and G.M. Ziegler, eds., vol. 29 of DMV Seminar, Basel: Birkhäuser, 2000, pp. 131–154. (Cited on p. 62)

[17] T. BUI-THANH AND M. GIROLAMI, *Solving Large-Scale PDE-Constrained Bayesian Inverse Problems with Riemann Manifold Hamiltonian Monte Carlo*, preprint, arXiv:1407.1517, 2014. (Cited on p. 66)

[18] T. BUI-THANH, K. WILLCOX, AND O. GHATTAS, *Model reduction for large-scale systems with high-dimensional parametric input space*, SIAM Journal on Scientific Computing, 30 (2008), pp. 3270–3288. (Cited on p. 16)

[19] H.-J. BUNGARTZ AND M. GRIEBEL, *Sparse grids*, Acta Numerica, 13 (2004), pp. 147–269. (Cited on p. 60)

[20] R.E. CAFLISCH, *Monte carlo and quasi-Monte Carlo methods*, Acta Numerica, 7 (1998), pp. 1–49. (Cited on pp. 61, 62)

[21] Y. CAO, S. LI, L. PETZOLD, AND R. SERBAN, *Adjoint sensitivity analysis for differential-algebraic equations: The adjoint DAE system and its numerical solution*, SIAM Journal on Scientific Computing, 24 (2003), pp. 1076–1089. (Cited on p. 14)

[22] K. CARLBERG, C. BOU-MOSLEH, AND C. FARHAT, *Efficient non-linear model reduction via a least-squares Petrov–Galerkin projection and compressive tensor approximations*, International Journal for Numerical Methods in Engineering, 86 (2011), pp. 155–181. (Cited on p. 16)

[23] K.S. CHAN AND C.J. GEYER, *Discussion: Markov chains for exploring posterior distributions*, The Annals of Statistics, 22 (1994), pp. 1747–1758. (Cited on pp. 59, 61)

[24] S. CHATURANTABUT AND D.C. SORENSEN, *Nonlinear model reduction via discrete empirical interpolation*, SIAM Journal on Scientific Computing, 32 (2010), pp. 2737–2764. (Cited on p. 16)

[25] C.P. CHEN AND D. LIU, *Numerical investigation of supersonic combustion of the HyShot II in the shock tunnel*, Journal of Aeronautics, Astronautics and Aviation, 43 (2011), pp. 119–128. (Cited on p. 18)

[26] L.H. CHEN, *An inequality for the multivariate normal distribution*, Journal of Multivariate Analysis, 12 (1982), pp. 306–315. (Cited on p. 50)

[27] S. CHIB AND E. GREENBERG, *Understanding the Metropolis-Hastings algorithm*, The American Statistician, 49 (1995), pp. 327–335. (Cited on p. 59)

[28] P.G. CONSTANTINE, E. DOW, AND Q. WANG, *Active subspace methods in theory and practice: Applications to kriging surfaces*, SIAM Journal on Scientific Computing, 36 (2014), pp. A1500–A1524. (Cited on pp. viii, ix, 10, 11, 43, 45, 49, 52, 56, 57, 58, 59)

[29] P. CONSTANTINE, M. EMORY, J. LARSSON, AND G. IACCARINO, *Exploiting Active Subspaces to Quantify Uncertainty in the Numerical Simulation of the Hyshot II Scramjet*, preprint, arXiv:1408.6269, 2014. (Cited on pp. ix, 72, 74)

[30] P. CONSTANTINE AND D. GLEICH, *Computing Active Subspaces*, preprint, arXiv:1408.0545, 2014. (Cited on pp. ix, 26)

[31] P. CONSTANTINE, D.F. GLEICH, AND G. IACCARINO, *Spectral methods for parameterized matrix equations*, SIAM Journal on Matrix Analysis and Applications, 31 (2010), pp. 2681–2699. (Cited on p. 14)

[32] P.G. CONSTANTINE AND Q. WANG, *Residual minimizing model interpolation for parameterized nonlinear dynamical systems*, SIAM Journal on Scientific Computing, 34 (2012), pp. A2118–A2144. (Cited on p. 16)

[33] P.G. CONSTANTINE, M.S. ELDRED, AND E.T. PHIPPS, *Sparse pseudospectral approximation method*, Computer Methods in Applied Mechanics and Engineering, 229/232 (2012), pp. 1–12. (Cited on p. 16)

[34] P.G. CONSTANTINE, D.F. GLEICH, Y. HOU, AND J. TEMPLETON, *Model reduction with MapReduce-enabled tall and skinny singular value decomposition*, SIAM Journal on Scientific Computing, 36 (2014), pp. 5166–5191. (Cited on p. 16)

[35] P. G. CONSTANTINE, Q. WANG, A. DOOSTAN, AND G. IACCARINO, *A surrogate-accelerated Bayesian inverse analysis of the HyShot II flight data*, in 52nd AIAA/ASME/ASCE Structures, Structural Dynamics and Materials Conference, Denver, CO, paper AIAA-2011-2037, 2011. (Cited on pp. vii, 19)

[36] P.G. CONSTANTINE, Q. WANG, AND G. IACCARINO, *A Method for Spatial Sensitivity Analysis*. Annual Brief, Center for Turbulence Research, Stanford, CA, 2012. (Cited on p. 19)

[37] P.G. CONSTANTINE, B. ZAHARATOS, AND M. CAMPANELLI, *Discovering an Active Subspace in a Single-Diode Solar Cell Model*, preprint, arXiv:1406.7607, 2014. (Cited on p. ix)

[38] R.D. COOK, *Regression Graphics: Ideas for Studying Regressions through Graphics*, Wiley Series in Probability and Statistics, vol. 482, New York: John Wiley & Sons, 2009. (Cited on pp. 3, 7, 18)

[39] R.D. COOK AND L. NI, *Sufficient dimension reduction via inverse regression*, Journal of the American Statistical Association, 100 (2005), pp. 410–428. (Cited on p. 18)

[40] R.D. COOK AND S. WEISBERG, *Comment*, Journal of the American Statistical Association, 86 (1991), pp. 328–332. (Cited on p. 18)

[41] T. CRESTAUX, O. LE MAÎTRE, AND J.-M. MARTINEZ, *Polynomial chaos expansion for sensitivity analysis*, Reliability Engineering & System Safety, 94 (2009), pp. 1161–1172. Special Issue on Sensitivity Analysis. (Cited on p. 17)

[42] T. CUI, J. MARTIN, Y.M. MARZOUK, A. SOLONEN, AND A. SPANTINI, *Likelihood-informed dimension reduction for nonlinear inverse problems*, Inverse Problems, 30 (2014), paper 114015. (Cited on pp. 66, 67)

[43] P.J. DAVIS AND P. RABINOWITZ, *Methods of Numerical Integration*, Boston, MA: Academic Press, 1984. (Cited on pp. 59, 61)

[44] M.K. DEB, I.M. BABUSKA, AND J.T. ODEN, *Solution of stochastic partial differential equations using Galerkin finite element techniques*, Computer Methods in Applied Mechanics and Engineering, 190 (2001), pp. 6359–6372. (Cited on p. 14)

[45] A. DOOSTAN AND G. IACCARINO, *A least-squares approximation of partial differential equations with high-dimensional random inputs*, Journal of Computational Physics, 228 (2009), pp. 4332–4345. (Cited on p. 16)

[46] A. DOOSTAN AND H. OWHADI, *A non-adapted sparse approximation of PDEs with stochastic inputs*, Journal of Computational Physics, 230 (2011), pp. 3015–3034. (Cited on p. 16)

[47] E. DOW AND Q. WANG, *Output based dimensionality reduction of geometric variability in compressor blades*, in 51st AIAA Aerospace Sciences Meeting including the New Horizons Forum and Aerospace Exposition, Dallas, TX, 2013, paper AIAA-2013-0420. (Cited on p. 18)

[48] Q. DU, V. FABER, AND M. GUNZBURGER, *Centroidal Voronoi tessellations: Applications and algorithms*, SIAM Review, 41 (1999), pp. 637–676. (Cited on p. 62)

[49] B. EFRON AND R.J. TIBSHIRANI, *An Introduction to the Bootstrap*, Boca Raton, FL: CRC Press, 1994. (Cited on pp. 15, 35, 74)

[50] G. FARIVAR AND B. ASAEI, *Photovoltaic module single diode model parameters extraction based on manufacturer data sheet parameters*, in 2010 IEEE International Conference on Power and Energy (PECon), Selangor, Malaysia, 2010, pp. 929–934. (Cited on p. 78)

[51] M. FORNASIER, K. SCHNASS, AND J. VYBIRAL, *Learning functions of few arbitrary linear parameters in high dimensions*, Foundations of Computational Mathematics, 12 (2012), pp. 229–262. (Cited on pp. 18, 25)

[52] P. FRAUENFELDER, C. SCHWAB, AND R.A. TODOR, *Finite elements for elliptic problems with stochastic coefficients*, Computer Methods in Applied Mechanics and Engineering, 194 (2005), pp. 205–228. Selected papers from the 11th Conference on The Mathematics of Finite Elements and Applications. (Cited on p. 14)

[53] K. FUKUDA, *From the zonotope construction to the Minkowski addition of convex polytopes*, Journal of Symbolic Computation, 38 (2004), pp. 1261–1272. (Cited on pp. 48, 62)

[54] K. FUKUMIZU AND C. LENG, *Gradient-based kernel dimension reduction for regression*, Journal of the American Statistical Association, 109 (2014), pp. 359–370. (Cited on p. 18)

[55] A. GARDNER, K. HANNEMANN, J. STEELANT, AND A. PAULL, *Ground testing of the HyShot supersonic combustion flight experiment in HEG and comparison with flight data*, in 40th AIAA Joint Propulsion Conference, American Institute of Aeronautics and Astronautics, Fort Lauderdale, FL, 2004, paper AIAA-2004-3345. (Cited on p. 71)

[56] M. GASCA AND T. SAUER, *Polynomial interpolation in several variables*, Advances in Computational Mathematics, 12 (2000), pp. 377–410. (Cited on p. 49)

[57] A. GENZ AND B.D. KEISTER, *Fully symmetric interpolatory rules for multiple integrals over infinite regions with Gaussian weight*, Journal of Computational and Applied Mathematics, 71 (1996), pp. 299–309. (Cited on p. 61)

[58] R.G. GHANEM AND P.D. SPANOS, *Stochastic Finite Elements: A Spectral Approach*, New York: Springer, 1991. (Cited on p. 14)

[59] A. GITTENS AND J.A. TROPP, *Tail Bounds for All Eigenvalues of a Sum of Random Matrices*, preprint, arXiv:1104.4513, 2011. (Cited on pp. 26, 27)

[60] G.H. GOLUB AND C.F. VAN LOAN, *Matrix Computations*, 3rd ed., Baltimore: Johns Hopkins University Press, 1996. (Cited on pp. 30, 32, 34)

[61] A. GRIEWANK, *Evaluating Derivatives: Principles and Techniques of Algorithmic Differentiation*, Philadelphia: SIAM, 2000. (Cited on p. 14)

[62] H. HAARIO, M. LAINE, A. MIRA, AND E. SAKSMAN, *DRAM: Efficient adaptive MCMC*, Statistics and Computing, 16 (2006), pp. 339–354. (Cited on p. 66)

[63] C.W. HANSEN, *Estimation of Parameters for Single Diode Models Using Measured IV Curves*, 39th IEEE Technical Report, Sandia National Laboratories, Albuquerque, NM, 2013. http://energy.sandia.gov/wp/wp-content/gallery/uploads/Hansen_SAND2013-4759C_PVSC391.pdf (Cited on p. 78)

[64] C.W. HANSEN, A. LUKETA-HANLIN, AND J.S. STEIN, *Sensitivity of single diode models for photovoltaic modules to method used for parameter estimation*, in 28th European Photovoltaic Solar Energy Conference and Exhibition, Paris, 2013, pp. 3258–3264. (Cited on p. 78)

[65] T. HASTIE, R. TIBSHIRANI, AND J. FRIEDMAN, *The Elements of Statistical Learning*, 2nd ed., Berlin: Springer, 2009. (Cited on p. 49)

[66] J.C. HELTON, J.D. JOHNSON, AND W.L. OBERKAMPF, *An exploration of alternative approaches to the representation of uncertainty in model predictions*, Reliability Engineering & System Safety, 85 (2004), pp. 39–71. Alternative Representations of Epistemic Uncertainty. (Cited on p. 14)

[67] F.J. HICKERNELL, *My dream quadrature rule*, Journal of Complexity, 19 (2003), pp. 420–427. Oberwolfach Special Issue. (Cited on p. 60)

[68] D. HIGDON, J. GATTIKER, B. WILLIAMS, AND M. RIGHTLEY, *Computer model calibration using high-dimensional output*, Journal of the American Statistical Association, 103 (2008), pp. 570–583. (Cited on p. 14)

[69] M. HRISTACHE, A. JUDITSKY, J. POLZEHL, AND V. SPOKOINY, *Structure adaptive approach for dimension reduction*, The Annals of Statistics, 29 (2001), pp. 1537–1566. (Cited on p. 18)

[70] G. IACCARINO, R. PECNIK, J. GLIMM, AND D. SHARP, *A QMU approach for characterizing the operability limits of air-breathing hypersonic vehicles*, Reliability Engineering and System Safety, 96 (2011), pp. 1150–1160. (Cited on p. 71)

[71] I. JOLLIFFE, *Principal Component Analysis*, New York: Springer, 2002. (Cited on pp. 35, 37)

[72] D.R. JONES, *A taxonomy of global optimization methods based on response surfaces*, Journal of Global Optimization, 21 (2001), pp. 345–383. (Cited on p. 73)

[73] J. KAIPIO AND E. SOMERSALO, *Statistical and Computational Inverse Problems*, New York: Springer, 2005. (Cited on pp. 65, 66, 67)

[74] M.C. KENNEDY AND A. O'HAGAN, *Bayesian calibration of computer models*, Journal of the Royal Statistical Society. Series B (Statistical Methodology), 63 (2001), pp. 425–464. (Cited on p. 14)

[75] G.K.W. KENWAY AND J.R.R.A. MARTINS, *Multi-point high-fidelity aerostructural optimization of a transport aircraft configuration*, Journal of Aircraft, 51 (2012) pp. 144–160. (Cited on p. 81)

[76] J.R. KOEHLER AND A.B. OWEN, *Computer experiments*, in Design and Analysis of Experiments, S. Ghosh and C.R. Rao, eds., vol. 13 of Handbook of Statistics, Amsterdam: Elsevier, 1996, pp. 261–308. (Cited on pp. 14, 59)

[77] S. KUCHERENKO, M. RODRIGUEZ-FERNANDEZ, C. PANTELIDES, AND N. SHAH, *Monte Carlo evaluation of derivative-based global sensitivity measures*, Reliability Engineering & System Safety, 94 (2009), pp. 1135–1148. Special Issue on Sensitivity Analysis. (Cited on pp. 17, 23)

[78] L. LE CAM AND G. LO YANG, *Asymptotics in Statistics: Some Basic Concepts*, New York: Springer, 2000. (Cited on p. 68)

[79] O.P. LE MAÎTRE AND O.M. KNIO, *Spectral Methods for Uncertainty Quantification: With Applications to Computational Fluid Dynamics*, Dordrecht, The Netherlands: Springer, 2010. (Cited on p. 14)

[80] B. LI, H. ZHA, AND F. CHIAROMONTE, *Contour regression: A general approach to dimension reduction*, The Annals of Statistics, 33 (2005), pp. 1580–1616. (Cited on p. 18)

[81] K.-C. LI, *Sliced inverse regression for dimension reduction*, Journal of the American Statistical Association, 86 (1991), pp. 316–327. (Cited on p. 18)

[82] ———, *On principal Hessian directions for data visualization and dimension reduction: Another application of Stein's lemma*, Journal of the American Statistical Association, 87 (1992), pp. 1025–1039. (Cited on p. 18)

[83] Q. LI AND J.S. RACINE, *Nonparametric Econometrics: Theory and Practice*, Princeton, NJ: Princeton University Press, 2007. (Cited on pp. 25, 49, 76)

[84] C. LIEBERMAN, K. WILLCOX, AND O. GHATTAS, *Parameter and state model reduction for large-scale statistical inverse problems*, SIAM Journal on Scientific Computing, 32 (2010), pp. 2523–2542. (Cited on p. 18)

[85] M. LOÈVE, *Probability Theory II*, Heidelberg: Springer-Verlag, 1978. (Cited on p. 42)

[86] L. LOVÁSZ AND M. SIMONOVITS, *Random walks in a convex body and an improved volume algorithm*, Random Structures & Algorithms, 4 (1993), pp. 359–412. (Cited on p. 62)

[87] T.W. LUKACZYK, F. PALACIOS, J.J. ALONSO, AND P. CONSTANTINE, *Active subspaces for shape optimization*, in 10th AIAA Multidisciplinary Design Optimization Conference, National Harbor, MD, 2014, paper AAIA-2014-1171. (Cited on pp. ix, 82, 84)

[88] J. MARTIN, L.C. WILCOX, C. BURSTEDDE, AND O. GHATTAS, *A stochastic Newton MCMC method for large-scale statistical inverse problems with application to seismic inversion*, SIAM Journal on Scientific Computing, 34 (2012), pp. A1460–A1487. (Cited on p. 66)

[89] H.G. MATTHIES AND A. KEESE, *Galerkin methods for linear and nonlinear elliptic stochastic partial differential equations*, Computer Methods in Applied Mechanics and Engineering, 194 (2005), pp. 1295–1331. Special Issue on Computational Methods in Stochastic Mechanics and Reliability Analysis. (Cited on p. 14)

[90] B. MOHAMMADI, *Uncertainty quantification by geometric characterization of sensitivity spaces*, Computer Methods in Applied Mechanics and Engineering, 280 (2014), pp. 197–221. (Cited on p. 18)

[91] J.J. MORÉ AND S.M. WILD, *Estimating derivatives of noisy simulations*, ACM Transactions on Mathematical Software, 38 (2012), paper 19. (Cited on pp. 14, 33)

[92] J.J. MORÉ AND S.M. WILD, *Do you trust derivatives or differences?*, Journal of Computational Physics, 273 (2014), pp. 268–277. (Cited on pp. 14, 33)

[93] M.D. MORRIS, *Factorial sampling plans for preliminary computational experiments*, Technometrics, 33 (1991), pp. 161–174. (Cited on p. 17)

[94] F. NOBILE, R. TEMPONE, AND C.G. WEBSTER, *A sparse grid stochastic collocation method for partial differential equations with random input data*, SIAM Journal on Numerical Analysis, 46 (2008), pp. 2309–2345. (Cited on p. 16)

[95]   J. NOCEDAL AND S.J. WRIGHT, *Numerical Optimization*, 2nd ed., Heidelberg: Springer, 2006. (Cited on p. 62)

[96]   E. NOVAK, K. RITTER, R. SCHMITT, AND A. STEINBAUER, *On an interpolatory method for high dimensional integration*, Journal of Computational and Applied Mathematics, 112 (1999), pp. 215–228. (Cited on p. 61)

[97]   A. O'HAGAN, *Bayes–Hermite quadrature*, Journal of Statistical Planning and Inference, 29 (1991), pp. 245–260. (Cited on p. 60)

[98]   I.V. OSELEDETS, *Tensor-train decomposition*, SIAM Journal on Scientific Computing, 33 (2011), pp. 2295–2317. (Cited on p. 16)

[99]   A.B. OWEN, *Variance components and generalized Sobol' indices*, SIAM/ASA Journal on Uncertainty Quantification, 1 (2013), pp. 19–41. (Cited on p. 17)

[100]   A.B. OWEN, *Monte Carlo Theory, Methods and Examples*, manuscript, 2013. http://www-stat.stanford.edu/~owen/mc. (Cited on pp. 5, 15, 52)

[101]   H. OWHADI, C. SCOVEL, T.J. SULLIVAN, M. MCKERNS, AND M. ORTIZ, *Optimal uncertainty quantification*, SIAM Review, 55 (2013), pp. 271–345. (Cited on p. 15)

[102]   F. PALACIOS, J.J. ALONSO, M. COLONNO, J. HICKEN, AND T. LUKACZYK, *Adjoint-based method for supersonic aircraft design using equivalent area distributions*, in 50th AIAA Aerospace Sciences Meeting including the New Horizons Forum and Aerospace Exposition, Nashville, TN, 2012, paper 2012-0269. (Cited on p. 81)

[103]   F. PALACIOS, M.R. COLONNO, A.C. ARANAKE, A. CAMPOS, S.R. COPELAND, T.D. ECONOMON, A.K. LONKAR, T.W. LUKACZYK, T.W.R. TAYLOR, AND J.J. ALONSO, *Stanford University unstructured (SU^2): An open source integrated computational environment for multiphysics simulation and design*, in 51st AIAA Aerospace Sciences Meeting and Exhibit, Grapevine, TX, 2013, paper AIAA-2013-287. (Cited on p. 82)

[104]   J. PENG, J. HAMPTON, AND A. DOOSTAN, *A weighted ℓ_1-minimization approach for sparse polynomial chaos expansions*, Journal of Computational Physics, 267 (2014), pp. 92–111. (Cited on p. 16)

[105]   E. PRULIERE, F. CHINESTA, AND A. AMMAR, *On the deterministic solution of multidimensional parametric models using the proper generalized decomposition*, Mathematics and Computers in Simulation, 81 (2010), pp. 791–810. (Cited on p. 16)

[106]   C.E. RASMUSSEN AND C.K. WILLIAMS, *Gaussian Processes for Machine Learning*, vol. 1, Cambridge, MA: MIT Press, 2006. (Cited on pp. 23, 57)

[107]   C.W. ROWLEY, *Model reduction for fluids, using balanced proper orthogonal decomposition*, International Journal of Bifurcation and Chaos in Applied Sciences and Engineering, 15 (2005), pp. 997–1013. (Cited on p. 16)

[108]   T.M. RUSSI, *Uncertainty Quantification with Experimental Data and Complex System Models*, Ph.D. thesis, University of California, Berkeley, 2010. (Cited on pp. vii, 18, 26)

[109]   J. SACKS, W.J. WELCH, T.J. MITCHELL, AND H.P. WYNN, *Design and analysis of computer experiments*, Statistical Science, 4 (1989), pp. 409–423. (Cited on p. 14)

[110]   A. SALTELLI, M. RATTO, T. ANDRES, F. CAMPOLONGO, J. CARIBONI, D. GATELLI, M. SAISANA, AND S. TARANTOLA, *Global Sensitivity Analysis: The Primer*, New York: John Wiley & Sons, 2008. (Cited on pp. 7, 17, 81)

[111] A.M. SAMAROV, *Exploring regression structure using nonparametric functional estimation*, Journal of the American Statistical Association, 88 (1993), pp. 836–847. (Cited on p. 23)

[112] R.C. SMITH, *Uncertainty Quantification: Theory, Implementation, and Applications*, Philadelphia: SIAM, 2013. (Cited on p. 13)

[113] I.M. SOBOL', *Multidimensional Quadrature Formulas and Haar Functions*, Moscow: Nauka, 1969. In Russian. (Cited on p. 17)

[114] I.M. SOBOL', *Global sensitivity indices for nonlinear mathematical models and their Monte Carlo estimates*, Mathematics and Computers in Simulation, 55 (2001), pp. 271–280. The Second IMACS Seminar on Monte Carlo Methods. (Cited on p. 17)

[115] I.M. SOBOL', S. TARANTOLA, D. GATELLI, S.S. KUCHERENKO, AND W. MAUNTZ, *Estimating the approximation error when fixing unessential factors in global sensitivity analysis*, Reliability Engineering & System Safety, 92 (2007), pp. 957–960. (Cited on p. 17)

[116] G. STEWART, *Error and perturbation bounds for subspaces associated with certain eigenvalue problems*, SIAM Review, 15 (1973), pp. 727–764. (Cited on pp. 32, 37)

[117] M. STOYANOV AND C.G. WEBSTER, *A gradient-based sampling approach for dimension reduction of partial differential equations with stochastic coefficients*, International Journal for Uncertainty Quantification, to appear. (Cited on p. 18)

[118] A.M. STUART, *Inverse problems: A Bayesian perspective*, Acta Numerica, 19 (2010), pp. 451–559. (Cited on pp. 15, 65)

[119] B. SUDRET, *Global sensitivity analysis using polynomial chaos expansions*, Reliability Engineering & System Safety, 93 (2008), pp. 964–979. Bayesian Networks in Dependability. (Cited on p. 17)

[120] R. TIPIREDDY AND R. GHANEM, *Basis adaptation in homogeneous chaos spaces*, Journal of Computational Physics, 259 (2014), pp. 304–317. (Cited on p. 18)

[121] J.A. TROPP, *User-friendly tail bounds for sums of random matrices*, Foundations of Computational Mathematics, 12 (2012), pp. 389–434. (Cited on pp. 26, 30)

[122] J.A. VRUGT, C.J.F. TER BRAAK, C.G.H. DIKS, B.A. ROBINSON, J.M. HYMAN, AND D. HIGDON, *Accelerating Markov chain Monte Carlo simulation by differential evolution with self-adaptive randomized subspace sampling*, International Journal of Nonlinear Sciences and Numerical Simulation, 10 (2009), pp. 273–290. (Cited on p. 66)

[123] H. WANG AND Y. XIA, *Sliced regression for dimension reduction*, Journal of the American Statistical Association, 103 (2008), pp. 811–821. (Cited on p. 18)

[124] H. WENDLAND, *Scattered Data Approximation*, vol. 2, Cambridge, UK: Cambridge University Press, 2005. (Cited on p. 49)

[125] S.M. WILD, R.G. REGIS, AND C.A. SHOEMAKER, *ORBIT: Optimization by radial basis function interpolation in trust-regions*, SIAM Journal on Scientific Computing, 30 (2008), pp. 3197–3219. (Cited on p. 73)

[126] D. WILLIAMS, *Probability with Martingales*, Cambridge, UK: Cambridge University Press, 1991. (Cited on pp. 49, 51, 60)

[127] Y. XIA, *A constructive approach to the estimation of dimension reduction directions*, The Annals of Statistics, 35 (2007), pp. 2654–2690. (Cited on p. 18)

[128] Y. XIA, H. TONG, W.K. LI, AND L.-X. ZHU, *An adaptive estimation of dimension reduction space*, Journal of the Royal Statistical Society: Series B (Statistical Methodology), 64 (2002), pp. 363–410. (Cited on p. 18)

[129] D. XIU AND J.S. HESTHAVEN, *High-order collocation methods for differential equations with random inputs*, SIAM Journal on Scientific Computing, 27 (2005), pp. 1118–1139. (Cited on p. 16)

[130] D. XIU AND G.E. KARNIADAKIS, *The Wiener–Askey polynomial chaos for stochastic differential equations*, SIAM Journal on Scientific Computing, 24 (2002), pp. 619–644. (Cited on p. 14)

Index